Voices from the Past, Visions for the Future

Voices from the Past, Visions for the Future:

A Modern Assessment of Harry F. Chaddick's 18 Essential Planning Ideas for Chicago

Martin E. Toth

Joseph P. Schwieterman, Ph.D., Editor

VOICES FROM THE PAST, VISIONS FOR THE FUTURE:
A Modern Assessment of Harry F. Chaddick's
18 Essential Planning Ideas for Chicago

International Standard Book Number: 0-9679061-0-5

Chaddick Institute for Metropolitan Development
DePaul University
243 S. Wabash Avenue, Suite 9000
Chicago, IL 60604

312/362-5731 (voice)
312/362-5506 (fax)

http://www.depaul.edu/~chaddick

The author wishes to thank the following individuals who provided editorial and technical assistance: Matthew Asciutto; Robert J. Boylan, AICP; Lara Duda; Laurie Marston, AICP; Jonathan Moore; and Richard Trevino.

Cover photograph credit: Peter J. Schulz/City of Chicago

Photographs on pages 30, 47, 52, 84, 92, and 100 appear courtesy of Joseph P. Schwieterman and Martin E. Toth.

Book design by Mainline Publications Inc.

Dedicated to the memory of
Camille Chaddick Hatzenbuehler
(1935–1999)

Camille Chaddick Hatzenbuehler

(1935–1999)

Foreword

This book is lovingly dedicated to the memory of Camille Chaddick Hatzenbuehler, the only child of Harry F. Chaddick.

While Harry Chaddick worked to preserve and improve the infrastructure of his beloved city of Chicago, Camille dedicated her life to making it aesthetically and artistically beautiful. Her tireless work for Chicago's civic treasures is legendary. She showed her love for the city's history by serving the Chicago Convention and Tourism Bureau, Know Your Chicago, and Chicago Sister Cities. The Lyric Opera of Chicago, the Chicago Symphony, and the Art Institute benefited from her passion for the arts. And her drive to improve the lives of all Chicagoans is still evident today thanks to her hard work for countless charities. Chicago is better for having such an extraordinary woman on its big shoulders.

Thank you, Camille, for teaching us that giving truly is better than receiving. The Chaddick Institute celebrates that legacy.

Elaine M. Chaddick
President
Harry F. and Elaine M. Chaddick Foundation

Contents

Introduction

Wisdom lies neither in fixity nor in change,
but in the dialectic between the two.
Octavio Paz

Noted entrepreneurs and influential civil servants often possess an innate ability to anticipate problems that might otherwise catch the general public by surprise. Such individuals have an eye for detail and a comprehension of the nuances of the market and political processes that prove elusive to others. They also are willing to defy conventional wisdom and pursue actions that initially may be received skeptically by the public at large.

Harry F. Chaddick had such talents. As a public servant, entrepreneur, and philanthropist, Chaddick left a profound imprint on the Chicago metropolitan area. Although many members of younger generations are unfamiliar with the contributions of Chaddick, who died in 1994, he should be remembered for his prominent role in transforming Chicago from a boisterous industrial city into the dynamic "city that works"—a place befitting its reputation as one of the world's truly great urban areas.

This book offers a retrospective look at 18 of the planning ideas Chaddick espoused throughout his long and illustrious career. It weighs the progress—or, in some cases, the lack of progress—that has been made on issues Chaddick considered essential to the betterment of our regional community. In reviewing notable successes and failures in Chicago's efforts to facilitate effective urban planning, salient lessons emerge from Chaddick's vision for its future.

These vignettes offer a glimpse of the formidable planning issues Chaddick faced in his years of public service. Each begins with a brief statement from Chaddick, taken from personal interviews, articulating his views and providing a series of recommendations for generations to follow. The essays conclude with a contemporary look at the status of Chaddick's many ideas in our evolving urban milieu.

From the West Side

Throughout his professional life, Chaddick stood on the shoulders of giants. He was born on Chicago's West Side in 1902 and grew up in modest surroundings. Like other visionaries from Illinois, Chaddick's earliest job portended a remarkable career. George Pullman built cabinets before building railway sleeping cars. Samuel Insull served as the personal secretary to Thomas Edison before establishing his empire in electric railways and utilities. Chaddick started as an errand boy in the Loop before rising to become a leader in the trucking industry, a major developer of commercial and industrial complexes, and a faithful public servant and trusted ally of Mayor Richard J. Daley.

Why should Chaddick be remembered? His path-breaking accomplishments spanned a half-century. They demonstrated his ability to make complex tasks simpler and more efficient. By his 35th birthday, Chaddick had helped refine a trucking innovation known as "piggybacking"—the shipment of truck trailers on railroad flatcars—which rescued the Chicago Great Western Railroad from bankruptcy. Within a year, piggyback trains 50 cars long were rumbling from the West Side to the Twin Cities under his direction—a noteworthy accomplishment for a man who was a student at the Illinois Institute of

Mayor Richard J. Daley (left) greets Harry and Elaine Chaddick.
(Credit: Harry F. and Elaine M. Chaddick Foundation)

Technology only a few years earlier. "We knew it would work mechanically," Chaddick recalled in his autobiography, *Chaddick: Success Against All Odds.* "A loaded trailer could be lifted by a crane onto a flatcar and then secured by chains."

And work it did. So revolutionary was this idea that other railroads, envious of the Great Western's success, temporarily expelled the carrier from their trade organization. Officials wary of change embarked on a frenzied, if futile, search for legal reasons to disallow Chaddick's innovation. Even Chaddick's supporters failed to anticipate that his truck-on-flatcar innovation would eventually help bring thousands of jobs to Illinois and position Chicago as a formidable "intermodal" hub that would attract more than twice the traffic of its largest competitor.

Chaddick's subsequent success revolved around establishing a comprehensive system of trucking terminals on Chicago's West Side. Together with his associate Phyllis Sutker, a prominent industrial real estate broker, he established an integrated system of terminals—properties that would fuel the growth of mid-American trucking. This network eventually included a preponderance of the motor freight terminals in the Chicago area. Indeed, over the next three decades, Chaddick was to become to Chicago's trucking industry what George Halas was to Chicago football—a colorful personality boldly leading Illinois through entire generations of change.

His next innovation took the industry almost completely by surprise. He installed two-way radios in motor vehicles, bringing a stodgy trucking industry into the modern telecommunications era. Nevertheless, Chaddick first had to convince a skeptical American Trucking Association that the technology would work and that the group should embark on a costly project with Motorola of Chicago to put such technology into wide-

The American Transportation Company was one of Chaddick's many successful business ventures. *(Credit: Harry F. and Elaine M. Chaddick Foundation)*

spread use. He persuaded them by setting a transmitter atop Chicago's Midwest Athletic Building on Madison Street, demonstrating that the technology would prove effective. So impressed were they with Chaddick's innovation that they publicly broadcast the first two-way radio conversation involving a trucking company at their 1946 convention at Chicago's Sherman House hotel.

The innovation, quite literally, brought the trucking business into the modern telecommunications era and rendered one of Chaddick's many trucking companies, Standard Freight Lines, an industry leader. In the process, Chaddick became a familiar name throughout the Chicago region and a widely respected national figure in the transportation field.

A New Direction

The public images of some of Chicago's leading innovators in the transportation arena eventually tarnished. Insull fled to Europe when his electric railroads went bankrupt after the 1929 stock market crash; Pullman encountered a bloody strike that came to exemplify the most troubling aspects of labor-management discord. Nevertheless, Chaddick remained successful and well regarded as his career progressed. Indeed, he is perhaps remembered most for his contributions as a civil servant working to promote the city's interests while continuing to excel as an entrepreneur.

This second phase of Chaddick's career began during the early 1950s when Mayor Richard J. Daley asked him to become Chicago's director of zoning—an unusual move considering that Chaddick's acumen and savvy rested solidly in entrepreneurship and the rough-and-tumble transportation business. Still, the mayor's brazen move soon paid dividends. Chaddick signed off on all major real estate developments in the city for the next decade. He guided the rezoning of the entire town in 1957 and championed the adoption of modern land-use standards. Perhaps no other individual besides Daley could claim more responsibility for helping Chicago earn its reputation during that era as an exceptionally well-run city. Indeed, Chaddick claimed (with perhaps only a bit of hyperbole) to be familiar with all 643,000 parcels of land in Chicago.

As an astute businessman, Chaddick instinctively knew that Chicago's future could not be solved with thoughtful urban planning alone. At a time when most people thought the future of retailing rested comfortably in downtown megastores such as Goldblatt's and Wieboldt's, Chaddick called for Chicago to encourage the development of automobile-friendly shopping malls within the city's borders, thus combating suburban migration. "The basic problem, as I saw it, was an unwillingness on the part of retailers to stick their necks out by being the first in the city to establish malls," he observed.

As a result, this leading civic figure, now 50 years old, decided to "gamble" his own money to facilitate change. He purchased in the early 1960s the old Ford plant off Cicero Avenue just south of Midway Airport. This mammoth facility, originally built to manufacture aircraft engines during World War II, was so unsightly that it was dubbed the "World's Largest White Elephant." Chaddick spent millions converting this beleaguered industrial facility into Chicago's first major shopping mall. Within a few months after its 1962 opening, he had convinced 82 retailers to sign leases and the U.S. Post Office to rent a half-million square feet in his nearby industrial park.

Soon, Ford City Shopping Center became a bustling commercial venue and a resounding success. Only a decade later, Ford City generated some 5,500 jobs; the nearby office park produced 20,000 more, bringing $12 million to Chicago in property taxes alone. And its prosperity would prove to be long-lasting. Although Chaddick sold the mall in the 1970s, the entire area would continue to enjoy robust growth. In fact, the shopping center became so important to Chicago's Southwest Side that it served as an impetus for the construction of the Chicago Transit Authority Orange Line to nearby Midway Airport.

The success of Ford City convinced Chaddick that Chicago should take the offensive and boldly challenge the suburbs for dominance in retail trade. He understood that maintaining the status quo was a fatal strategy. Risking his reputation, he called for the city to open a series of seven regional shopping malls (with 150 or more stores) and 15 smaller community shopping centers on vacant or underdeveloped land, mostly abandoned by railroad companies.

With the valuable assistance of his long-time associates Robert J. Boylan, a certified urban planner, and John P. Murray, a real estate and zoning attorney, Chaddick dutifully prepared reports and studies to demonstrate the feasibility of his plan. Nevertheless, this grand vision proved to be beyond his considerable grasp. Chaddick could not overcome the entrenched resistance from local retailers who opposed the idea of new shopping facilities built in their backyards. Apart from downtown facilities such as Water Tower Place, only one more regional mall, the Brickyard, would be built, opening in 1973, and one community shopping center, Addison Mall, opening in 1984.

In retrospect, one can persuasively argue that Chaddick was among the few to anticipate Chicago's need to share in the commercial benefits of the growth of shopping malls. By 1973, there were 15 major regional shopping centers in Northeastern Illinois, but only one—Ford City—was within the city's borders. The age of urban decentralization had begun, and Chicago faced an enormous financial loss from the relocation of retail activity. Indeed, within a few years, millions of Chicago residents regularly drove out of town in search of modern shopping facilities. This problem evidently contributed (if perhaps only secondarily) to the migration of hundreds of thousands of residents out of the city.

In his characteristic businesslike fashion, Chaddick remained undaunted, rebounding from his only major setback. He turned his attention to developing real estate in Palm Springs, Calif.—a city that he helped transform from an inhospitable desert into a lush vacation mecca. In 1973, shortly after his 71st birthday, he opened his masterpiece, the 150-room Tennis Club Hotel, which would soon become the focal point of Palm Springs social life. Among its regular customers were numerous politicians and celebrities, including Red Skelton—a man whom Chaddick and his wife, Elaine, greatly admired.

Chaddick never forgot Chicago. He planted his fortunes back in the city he loved through a charitable organization, the Harry F. and Elaine M. Chaddick Foundation, which he co-chaired with his wife. Thus, the man who wore so many hats throughout his life ended his career by wearing the stately top hat of a philanthropist, donating money to hospitals, universities, and private charities.

A Fitting Tribute

There may be no better tribute to Harry F. Chaddick than to take a front-row seat on the Orange Line and witness his Chicago pass before you. On this trip, you will see the train filling with passengers at the Midway Airport station, including shoppers from the Ford City Shopping Center. As the train winds its way through the industrial Southwest Side, it passes not far from the old freight yards where Chaddick brought piggybacking to modern railroading. Entering Bridgeport, it passes near tranquil residential neighborhoods where one can see the fruits of Chaddick's decade-long effort to bring modern zoning practices to Chicago. A few blocks further east, as the train approaches Halsted Street, one of Chaddick's many trucking terminals from years ago becomes visible. Entering the Loop, crossing Madison, the train passes near the site where his two-way transmitters changed the face of trucking.

But it is the final stop—at Adams and Wabash—where the man's unassuming legacy is abundantly evident. Here, students can be seen heading to class at DePaul University's downtown campus to learn about urban planning and land-use as well as the virtues of public service—all because of a gift from Chaddick to the school seeking to perpetuate our city's rich tradition of civic service.

It was the final expression of Chaddick's deep conviction that everyone should understand—and ultimately participate in—the planning of local land-use and infrastructure throughout Illinois. He was concerned that citizens might leave all of the decisions in the hands of bureaucrats and politicians—a grave mistake in his view. Chaddick yearned for a chance to give citizens the opportunity to study regional economic issues and to have a significant impact on the planning and policy process. As he noted in his autobiography, "Cities need young men and women who have been schooled in the uses of urban land."

So committed was he to this goal that he boldly supported sometimes controversial positions—a fact evident in the remainder of this volume. "How can a city like Chicago continue to entrust its zoning and building decisions to aldermen who know little about either and have little or no information available to them," he lamented. Those that knew Chaddick, however, understood his intent was not simply to criticize. Indeed, in 1993, at the age of 91, Chaddick made one final investment, offering a donation to DePaul University to create the Chaddick Institute for Metropolitan Development.

The Institute focuses on the study of urban infrastructure, transportation, and land-use issues. It offers planning-related classes with DePaul's Public Services Graduate Program, sponsors professional conferences, hosts an annual technical workshop series, and coordinates a diverse range of research on contemporary urban themes. The Institute, while only a part of the larger world of academic and public policy organizations, reaches out to those from all walks of life, spreading the word to the community about the need for better urban planning and development.

If Chaddick's career was filled with paradox, it is fitting that many of the Institute's activities would take place at the DePaul Center, in the same building that once housed Goldblatt's department store. Ironically, this former bustling retail location lacked many modern conveniences and failed for precisely the reasons that Chaddick anticipated.

The passing of Chaddick and other civil servants of his era marks the end of more than a century of monumental public figures, including real estate developer Philip M. Klutznick and Mayor Richard J. Daley, who helped build modern-day Chicago. Perhaps a new generation of visionaries is preparing to carry on the torch of Illinois' luminaries and Chaddick's contemporaries. But it is more likely that major corporations, led by less memorable figures, will be steering us in the decades ahead.

Regional Needs

1

Waterway Transportation

Improvement and Coordination of Passenger Transportation over Regional Waterways

"Ideally, buses providing transportation for passengers from the suburbs or neighborhoods near the city's limits should travel on unobstructed expressways. In Chicago, there are some unique potential 'expressways' which should be considered as part of our transportation system. These expressways are components of the city's inland waterway system. Fortunately, for transportation purposes, each of these waterways leads to the downtown area. All should be examined for potential use by state-of-the-art watercraft.

"The potential for commuter transportation on our nation's waterways appears to have been ignored by most city and transportation planners. At the present time, the only commuter services utilizing a water transit mode in the United States are ferry services, which link cities or sections of cities. The first of our waterways that might be used for passenger service is the Chicago Sanitary and Ship Canal, which runs through southwest suburbs to connect with DuPage County, as well as the Calumet River, Lake Michigan, and downtown Chicago. The second waterway is the north branch of the Chicago River, leading directly to the Loop from as far north as perhaps the North Shore suburbs. The third waterway is the lake itself, which could make waterborne service possible for the north and south sides, and the entire south end of Lake Michigan.

"Experimentation should begin now to determine the feasibility of adding high-speed watercraft (including, perhaps, hydrofoil or air-cushion) to the Chicago metro-area transportation system."

— Harry F. Chaddick

Contemporary Assessment

With growing numbers of cities turning to their waterways for transit service in addition to freight transport, Chaddick's call for heightened emphasis on waterborne transportation remains both timely and relevant. With smaller-scale water taxi service already in place, planners in Chicago are rediscovering this long-neglected form of conveyance, which promises to offer an additional method of reducing automobile use and alleviating lakefront corridor congestion. The continuing downtown renaissance portends a promising future for marine-oriented transit.

The following paragraphs summarize essential issues surrounding transit service on Chicago's waterways.

The waterway transit industry, consisting predominantly of local and interurban ferry service, has made dramatic gains in both ridership and municipal investment.

Several U.S. cities notable for diverse maritime histories, including Boston, San Francisco, and New York, have witnessed a surge in ferry ridership in recent years. So, too, has Washington, D.C., where taxi service has become a popular transportation option along the Potomac River. In these and other places, waterborne transit is reclaiming its position as an integral part of the regional transportation network. Such a resurgence is attributable to both rising demand and large municipal investments to improve and expand terminals as well as acquire new watercraft. Recognizing that many waterways are little used even during the busiest travel periods, regional planners are embracing them as an expedient means of fostering urban mobility.

Although federal authorities do not compile comprehensive statistics for the waterway transit sector, reputable estimates suggest that urban ferry ridership more than doubled between 1986 and 1996, reaching about 60 million trips annually (not including New York's Staten Island Ferry with its relatively stable daily average of 60,000 passengers). This recent rise in ridership has forced many transportation officials to reconsider the traditional "rubber-tire mentality" that has historically favored cars and buses.

Several proposals have been made in recent years to reinstate ferry service across lower Lake Michigan, but all have failed to acquire sufficient financial backing. One company's proposal would initially provide commuter service to Chicago from Michigan City, Ind., and New Buffalo, Mich., and eventually expand its network to include Milwaukee, Wisc., and Gary, Hammond, and East Chicago, Ind. Seeking to reduce congestion along the Gary-Chicago-Milwaukee corridor, another recent proposal would establish truck and auto-passenger ferries operating between Milwaukee, Wisc., and a Michigan port, bypassing Chicago completely.

Local tour and cruise ship operators are finding water taxi service a lucrative market.

Water-based transportation has been a diminutive part of the urban transportation picture in Chicago over much of the past century, with only Wendella Sightseeing Boats providing continuous service since the 1930s. In response to the renovation of Navy Pier and other lakefront improvement projects, however, the new millennium brings the promise of an enormous rise in tourism and convention business on Lake Michigan and the Chicago River. In the process, several water taxi operators are creating a market niche by simultaneously conveying both commuters and tourists to destinations along the river and lake.

Wendella remained the sole provider of a true commuter service for many years, shuttling passengers between its dock at

A water taxi navigates the Chicago River near the State Street bridge.
(Credit: Javet M. Kimble/City of Chicago)

Madison Street near the Ogilvie Transportation Center (formerly the Chicago and North Western station) to just below the Wrigley Building on Michigan Avenue. During the past two years, however, Wendella extended its schedule beyond rush hour and three new water taxi services started operating, each offering commuters and tourists various destinations and types of service. One of its competitors, Shoreline Sightseeing, operates all day on Lake Michigan between Navy Pier and the Shedd Aquarium, and from Navy Pier to the Sears Tower. Another provider, Chicago Fire Boat Cruise Co., shuttles passengers between the Ogilvie Transportation Center and the base of the IBM Building at Wabash Avenue. The industry's newest entrant, Chicago From the Lake, sails from the often-crowded docks at Navy Pier. This company's smaller boats provide a more flexible service and are able to pick up and drop passengers where they choose—provided the boat acquires permission to dock.

The expansion from one to four operators in only two years exemplifies the growing interest among entrepreneurs in providing water transportation to commuters weary of using crowded arterial roads. While Chicago's weather prevents commuter boats from operating during the winter months, one could easily imagine such services operating on regular schedules between public transit stations and the expanding commercial area north of the Chicago River during warmer seasons.

The city of Chicago has only recently begun to promote water taxi service as an alternative to help alleviate automobile traffic congestion.

After long being oblivious to (and in many cases skeptical of) the role of local waterways as transit resources, planners began in 1993 to reassess water taxi service. Initially, officials considered licensing and rental fees, as well as a tax on ticket sales as possible revenue sources to help offset costs of riverfront beautification projects.

The city then opened the operating license availability to competitive bidding, and the immediate interest from private tour companies clearly showed the extent of demand in this market. The significant response to the bidding, along with interest from local property owners, prompted the city to authorize the Departments of Revenue and Consumer Services to draft regulations to oversee taxi services, routes, and fares. However, the city's formal authority effectively ends with the licensing process largely because the federal government has jurisdiction over Lake Michigan and the Chicago River.

As a result of robust Loop-area expansion, municipal officials and industry analysts predict long-term growth for the water taxi industry. The principal obstacle in the interim will be convincing owners of riverfront property to grant docking rights to these private carriers—a task which thus far has been difficult.

Taken in their entirety, the recent developments in waterway transit offer planners an additional option in meeting the ultimate goal of improved mobility. With more than 40 miles of navigable waterways in the city available for transit use, the ideas endorsed by Chaddick are gaining the favor of state and local policymakers. Considering the essential role rivers and waterways have played in the evolution of the Chicago region, it seems logical that we reclaim these public resources as avenues of passenger transportation.

2

Flood Prevention Systems

Definitive Study and Development of Appropriate Systems to Combat Recurring Flooding in the Chicago Metro Area

"Federal, state, and local taxpayers have spent $1.5 billion for construction of the Deep Tunnel system in Chicago to serve partly as a gigantic retaining basin in times of heavy rain. Nevertheless, at least twice a year the partially completed tunnel is filled to capacity. Polluted water must then be released into the Chicago River and then, eventually, to Lake Michigan, the source of the city's water supply. This is exactly what happened every year before the $1.5 billion was spent to build the tunnel.

"Major flooding, once or twice a year, continues to beset much of suburban Chicago, and the city itself. The primary cause is the lack of sufficient open and unpaved space for rainwater absorption and the thoughtless development changes to natural drainageways. Poorly regulated construction has proceeded for decades, covering a huge percentage of the land with concrete and asphalt. Little effort has been made to take into account land and water levels as well as retention capabilities. Residential development continues to proceed throughout northeastern Illinois with, in most instances, a lack of governmental stormwater retention requirements and commitments from real estate developers.

"To remedy this, legislation should be enacted that insists appropriate stormwater retention is provided and land be so graded as to allow water to flow naturally to nearby swales and drainage systems. Each day more and more land is being developed, but it is never too late to start doing what makes sense.

"It is certainly time for a regional agency to conduct a detailed engineering study of the metro area water drainage basin and develop a comprehensive program and a priority list of actions necessary to alleviate our continuing flooding hazards."

— Harry F. Chaddick

Contemporary Assessment

The short-term benefits of our region's massive investment in flood control may have been disappointing, as Chaddick suggests, but he would likely admire our diligence in seeing these projects through to completion—even though eradication of flood problems is still years away due to enormous engineering requirements. He might have initially expressed reservations about the budgetary considerations of the monumental Deep Tunnel project and other prolonged efforts to relieve flooding, but he would commend our methodical pursuit of these projects in the face of adversity.

A review of this issue suggests that several regional accomplishments deserve special consideration.

The immense Deep Tunnel project promises, when finally completed, to provide adequate drainage for Chicago and Cook County suburbs.

Deep Tunnel is part of the Tunnel and Reservoir Plan, whose design dates back to the 1960s. Although construction began during America's bicentennial year, the enormity of this project will prevent its completion until at least 2014. The huge undertaking is intended to protect our drinking water source, Lake Michigan, from pollution and help alleviate recurrent flooding throughout the region. One of the largest public works projects in our nation, the Tunnel and Reservoir Plan eventually will encompass a 109-mile system of tunnels connecting three

Movable bridges span the Chicago River with Lake Michigan in the distance.
(Credit: Peter J. Schulz/City of Chicago)

voluminous reservoirs and will provide a total capacity of 19 billion gallons for excess storm water and sewage—possibly expanding to 46 billion gallons by 2060.

When the massive Deep Tunnel network and pumping station apparatus is complete, nearly 4 million people will reap the benefits, and hundreds of thousands of basements will be less susceptible to flooding. The project resulted from devastating flooding in the region, including flooding from the Chicago River, which seriously damaged parts of downtown in 1954 and 1957. In the 1960s, officials hoped the Tunnel and Reservoir Plan would provide a definitive answer.

Only a small portion of those benefiting from Deep Tunnel truly understand how flood prevention in our region works. Over the years, Chicago and most older suburbs have combined sewers that carry both storm water and sewage. When heavy rains exceed the standard capacity of an inch per hour for local sewers, flooding and basement back-ups often result. The Metropolitan Water Reclamation District of Greater Chicago is able to divert water from overloaded sewers to Deep Tunnel, where it is stored and treated after the storm. Nevertheless, the tunnel system was not designed for storage and quickly fills, forcing officials to allow untreated storm water and sewage to flow directly into local rivers, canals, and Lake Michigan.

This condition will change markedly when a system of three reservoirs is complete. The smallest of the three reservoirs, the O'Hare Reservoir, began operating in 1998 and is expected to prevent more than $2.3 million per year in damage. This 350 million-gallon reservoir serves 21,000 homes and businesses in the northwest suburbs. The two much larger reservoirs will be located in south suburban McCook and Thornton. The Metropolitan Water Reclamation District of Greater Chicago has secured federal and local funding approval and will convert two gravel quarries into mammoth holding tanks for storm water and sewage. The McCook facility's first phase will provide a capacity of 10.5 billion gallons, while the Thornton reservoir will hold 9 billion gallons.

The construction of the tunnel system, thus far, has abetted efforts to clean up local waterways, notably the Chicago and Des Plaines Rivers, by absorbing the initial flush of heavily polluted water that accompanies a storm. To date, Deep Tunnel has served as a mini-reservoir, reducing the release of sewage-laden water into Lake Michigan from what was an almost weekly occurrence to an average of once a year.

Recent measures provide renewed hope for local flood alleviation.

Widespread flooding in 1997 provided the impetus for renewed public and governmental attention to flood control. In 1998, the South Suburban Mayors and Managers Association released a storm water strategy plan detailing how increased communication between communities can help limit flooding. By coordinating the design of flood control projects for participating communities, helping them implement warning systems, and establishing a broad model storm water ordinance, the association is attempting to show how a comprehensive approach is the best method to fight flooding and keep towns from simply diverting water—and hence the damages—to neighboring communities with lower elevations. Not only have communities such as Summit and River Grove recently embarked on multimillion-dollar projects to connect with Deep Tunnel, several other local storm drainage initiatives have also received municipal approval, such as a new pipeline project in DuPage County and a series of small projects in Downers Grove. Also in 1998, officials released a revised flood map, dramatically affecting building restrictions on floodplains across our region.

Several other measures contribute to the sense of optimism that prevails among flood control advocates. In 1998, in response to persistent resident requests, Chicago officials announced a $64 million pilot project to help alleviate basement flooding in neighborhoods across the city. A major part of this project is the installation of "vortex valves" (devices fitted with swirl chambers to regulate the flow of water) in catch basins throughout 17 wards, helping contain water on the street and keeping sewage from backing up into basements. Other important components of this project include a public education campaign to show homeowners how to disconnect downspouts from sewer lines to allow rain water to flow onto lawns, and a measure to increase the number of the city's relief sewers.

Developers and municipal officials are using enhanced development control measures to mitigate flooding.

As suburbs grow, homes and businesses continue to be built on floodplains, exposing our region's residents to greater cost and likelihood of flooding. Such development leads to widening floodplains and exposes still more buildings to potential flooding. Adding to this already vexing problem is the fact that only a quarter of all floodplain residents have federally subsidized flood insurance, thereby creating a burden for taxpayers when homeowners seek remuneration due to the effects of flooding. Although the outward migration of population will almost certainly continue, worsening area flooding, several communities have begun to promote development that respects wetlands, swales, and other natural methods of holding floodwater. Long Grove and Grayslake, for example, have formalized these techniques and consider them part of the community's overall development plans. (Notably, Chaddick pioneered the concept of sub-development retention basins by incorporating underground stormwater retention areas in his shopping center developments of the 1960s and 1970s.)

While officials and residents may be tempted to criticize Deep Tunnel for failing to prevent flooding after more than 20 years of construction, it should be remembered that the project has only recently begun the second of three development phases. This project bears testimony to the diligence of planners who have pursued innovative solutions to a difficult problem while receiving little public recognition for their efforts. Nevertheless, officials from the Metropolitan Water Reclamation District of Greater Chicago have acknowledged the need to improve public relations.

We can take pride in the progress that has been made throughout the region to integrate Deep Tunnel into a diverse, comprehensive flood-control plan that holds the promise of serving area residents for generations to come.

3

Environmentally Deficient Land

A Comprehensive Program at All Government Levels to Deal with Land Compromised by Toxic or Hazardous Materials

"My experience as a real estate developer and builder of shopping centers and industrial parks, in many respects, qualifies me to comment on the difficulties encountered when trying to improve land that has been tainted by previous exposure to toxic or hazardous elements, whether caused by nature or humans. Many developers have encountered frustrating delays and sizable financial losses (or given up completely) in their attempts to salvage and improve tainted land or—in some cases—land merely perceived to be tainted. The principal problems stem from lack of information, confusing or contradictory regulations, and lengthy paperwork and processing. Add to that mix trying to deal with multiple jurisdictions (federal, state, county, and local) and you produce an indigestible stew.

"It would seem that a coordinated program of recycling questionable land should begin with a careful survey and evaluation of all potentially tainted sites, with each site then 'rated' as to its development potential or relative ease of reclamation. This inventory should be made available to all planning and development agencies and to the development industry.

"Another major step forward would be a vigorous effort to have the various environmental enforcement jurisdictions join forces to coordinate their programs regarding 'damaged' lands so as to allow for the clarification and streamlining of regulations and acceleration of the review and approval process. The regulatory agencies might also

consider offering some form of bonus to developers willing to bear the cost of sanitizing a tainted site."

— Harry F. Chaddick

Contemporary Assessment

Chaddick witnessed firsthand the tumultuous events surrounding the decline of the manufacturing sector throughout the Midwest. More than a century of industrial development, often progressing with little regard for the environment, has left an indelible mark on the regional landscape. Although the Great Lakes region has gradually relinquished its image as America's "Rust Belt," many of its central cities remain permeated with abandoned and blighted properties—brownfields—that hamper redevelopment and restrict the expansion of adjacent industries.

The following considerations illustrate the profound importance of this issue.

The redevelopment of environmentally contaminated properties presents an enormous challenge to proponents of urban revitalization.

Although many foundries, factories, mills, and other symbols of America's industrial heritage have surrendered their role as contributors to the economy, they have left behind structures and storage areas that cover large portions of our major cities. As manufacturing companies closed in response to changes in market needs and technology, our cities became home to a plethora of properties now known as brownfields. These contaminated properties offer grim and potentially dangerous reminders of the consequences of the legacy of heavy industry in many neighborhoods.

The U.S. Environmental Protection Agency (EPA) defines brownfields as "abandoned, idled or underused industrial and commercial facilities where expansion or redevelopment is complicated by real or perceived environmental contamination." Based on this guideline, brownfield sites can range from parcels encompassing former gas stations or dry cleaners to large factories or railyards. Although the extent of the problem is difficult to assess, the U.S. General Accounting Office estimates there could be as many as 500,000 such sites nationwide, possibly requiring as much as $650 billion to clean them.

A former industrial site along Roosevelt Road adjacent to the South Branch of the Chicago River sits idle.

While many government officials and agencies have invested significant amounts of time and effort into promoting the benefits of redeveloping brownfields, businesses often look first to "greenfields"—undeveloped sites on the edge of or outside the urban area. Although offering spacious surroundings and convenient access to suburbs, the development of greenfields often necessitates the construction of costly infrastructure. Accordingly, this type of development can negatively affect the urban center through a loss of jobs, reduction of the tax base, diminished neighborhood vitality, and increased decentralization. A Conference of Mayors study of 100 cities from 1998 estimated that returning brownfields to productive use would add $200 million to $500 million to local economies in tax revenue alone and provide approximately 236,000 jobs.

Municipal officials throughout the nation recognize the opportunity cost of allowing brownfields to remain idle.

With a credible body of evidence indicating that many brownfields can be restored fairly inexpensively, a movement to return these sites to profitable use is gathering momentum. There is a growing belief among planners that brownfield redevelopment is a fiscally logical method of revitalizing inner cities. The public's rising awareness of environmental health issues is lending political support to this campaign.

Many municipal officials are discovering that brownfields are often not as contaminated as previously believed. In turn, federal, state, and local legislatures have worked since 1993 with the EPA and other agencies to enact laws to promote brownfield cleanup and redevelopment. The EPA's national brownfield program—an initiative lauded by both environmentalists and commercial organizations—chose 157 pilot areas between 1995 and 1998 that have leveraged over $1 billion in private investment. The agency added another 75 sites in 1999. More than 36 states, including Illinois, have adopted voluntary cleanup programs to address many of the prevailing disincentives to investing in brownfield sites.

Significant financial barriers pose continuing threats to the rehabilitation of brownfield sites.

Environmental and liability concerns remain the principal impediments to widespread brownfield redevelopment. Much of the recent progress in brownfield cleanup can be traced to the Comprehensive Environmental Response, Compensation, and Liability Act of 1980, generally referred to as "Superfund." This seminal piece of legislation removed much of the ambiguity surrounding responsibility for environmental remediation at some of the nation's most polluted sites. Superfund holds current and historical owners of property responsible for a substantial share of the cleanup costs and requires the property be returned to a pristine condition—regardless of the proposed use of that property. To encourage the timely completion of these projects, governments agencies have put into place a complex system of financial incentives.

While the legacy of Superfund remains unclear, the practice of holding current property owners liable for cleanup costs has clearly been a mixed blessing for older industrial cities. It has rendered many lenders and developers hesitant to invest in potentially contaminated property out of fear that they may inherit an enormous environmental liability. Although federal guidelines are gradually being amended to encourage brownfield redevelopment, few properties are sufficiently contaminated to be designated Superfund sites. Without strong financial incentives, therefore, many developers are resolute in their conviction that investments in brownfields are exceedingly risky.

Even with the recent success of brownfield redevelopment initiatives across the country, industrial expansion remains largely concentrated in greenfields. Rapidly expanding industries, such as pharmaceutical and computer hardware companies, need large tracts of land and proximity to a well-educated workforce.

The sprawling development pattern of many suburban communities often can most readily accommodate vast corporate campuses and factory complexes. With abundant space, lower property taxes, and, in some cases, less rigorous land-use controls, greenfields have remained the location of choice for many businesses.

With few federally designated Superfund sites in most cities and none in Chicago, lower units of government must accept much of the burden of cleanup. Many states and municipalities are working with the EPA to revise restrictive regulations and create initiatives that encourage businesses to redevelop brownfields. City officials tout the benefits of affordable cleanup costs, growing tax incentives, strategic location, transportation access, and a nearby workforce. By offering a diverse package of benefits, many cities are able to persuade businesses to resist the temptation to move to more pristine suburban areas.

Chicago deserves a great deal of credit for its pioneering role in promoting brownfield redevelopment.

The city's brownfield initiative, created in 1993 with the EPA and the Housing and Urban Development Department, has spent $1.3 million to acquire five sites deemed suitable for redevelopment. Other successful projects include the rehabilitation of a former bus barn facility and consequent sale of the land to a meat processing plant (which then hired an additional 150 employees) and the precedent-setting voluntary effort to reverse environmental damage at the enormous South Works site. The success of these efforts is providing the impetus to reclaim other parcels among an estimated 2,000 local brownfield sites.

Chicago has actively participated in national programs and supported local efforts that build on successful projects. The city earned the distinction of being named one of the nation's 16 Brownfield Showcase Communities in 1998, thereby attracting new funding and in-kind services from the recently established Brownfields National Partnership. This partnership coordinates the disparate resources of 17 federal agencies and thus promotes the efficient allocation of resources aimed at brownfield cleanup. By virtue of its designation as a federal empowerment zone, Chicago can also offer additional tax incentives to spur brownfield redevelopment. As of 1998, the city leveraged more than $56 million from public and private sources and received more funding in loan guarantees from local and federal agencies.

Chicago, like many other cities across the nation that have successfully promoted brownfield redevelopment, has gradually modified its strategies as it learns more about the costs and benefits of the cleanup process. Recognizing that contamination is often the least expensive part of revitalizing a site, officials have focused on removing other barriers to revitalization. A necessary first step is mitigating public fears about a site's possible contamination, while working simultaneously to overcome infrastructure, transportation, and land-assembly issues. When able to combine several smaller tracts into larger industrial parks and arrange investment to improve truck corridor access, city officials find they can effectively compete with greenfield sites. These newly developed industrial parks are then promoted strongly to small- and medium-sized firms—institutions that federal studies show to have the highest potential for new job creation.

While Chicago's accomplishments are diverse and impressive, one must acknowledge that its cleanup process remains in an embryonic state. Related to this, it deserves mention that some environmental groups are skeptical of the state's multitiered cleanup process, which establishes less stringent standards for land that will remain in industrial use. Others object to reforms in liability laws, fearful that they will allow major corporations to abrogate their responsibility for environmental remediation. Equally disconcerting is the lack of evidence showing that the city's cleanup efforts have significantly reduced the level of greenfield development in our region.

Although repairing our region's environmentally contaminated land is a task that will require the efforts of future generations, Chaddick would applaud the many remediation projects that have, in the opinion of some, made Chicago a standard-bearer for the national initiative.

4

Urban and Suburban Rail Transportation

An Effective System of Modern, Safe, and Fast Rail Transportation Linking the Region's Urban Centers

"Most urban studies show that residency follows jobs. As more and more job opportunities shift from Chicago to its suburbs and nearby satellite cities, more and more workers and their families will feel forced to move closer to their jobs. One of the ways our region must work to stem this flow (or to accommodate it less painfully) is to continually maintain and upgrade our transportation system.

"It is absolutely essential to talk about the need for new rail transportation, whether this takes the form of massive infrastructure improvements to the present system to accommodate newer and faster trains, or entirely new rail forms (such as monorail or magnetic-levitation devices).

"I am convinced that travel by modern rail (especially in the Midwest) between points less distant than 300 miles is, by far, the superior travel mode. Rail travel is less affected by weather and competing surface vehicles and is considerably safer.

"If fast, safe, and efficient rail service can be achieved in Europe and our own East Coast (New York to Washington, D.C.), why is the Chicago region content to accept slow, dangerous, and inefficient service?"

— Harry F. Chaddick

Contemporary Assessment

Chaddick would presumably be pleased with our region's progress in this area. Considerable strides have been made in evaluating, planning, and implementing several rail transportation projects—though, due to funding limitations, several of the more prominent proposals remain only in the design stage.

The paragraphs below summarize several prominent facets of this continuing discussion.

The demand for travel within the region is increasing at a rate faster than the rate of population growth.

The Chicago region is home to the nation's freight hub and the second-largest U.S. transit system. Its economy will remain closely tied to successfully maintaining and expanding its rail transportation system. The region's population is expected to increase from 7 million to 9 million and employment is expected to grow from 4 million to 5.5 million by 2020. As residents and businesses respond to this change, the demand for travel is projected to grow by 25 percent, and total vehicle-miles traveled is expected to rise at an even faster rate. Many expect regional freight movements to expand in similar proportion.

After two decades of decline, transit ridership is poised to continue the recent trend toward gradual growth.

Suffering from many years of indifferent service, rising fares, and the relocation of jobs and people away from Chicago and into the suburbs, transit is today in the midst of a modest renaissance. Metra ridership has risen steadily, setting a record for the second straight year in 1998 with more than 77 million passengers—a 3 percent gain over 1997. Chicago Transit Authority (CTA) ridership has also stabilized after a tumultuous period of decline. Moreover, the specter of worsening highway congestion, the aging of the population, and Illinois' recent commitment to

A Chicago Transit Authority Blue Line elevated train moves toward O'Hare International Airport. *(Credit: Peter J. Schulz/City of Chicago)*

increased funding for public transportation improvements will likely stimulate the demand for transit usage. The success of Metra's North Central commuter rail line to Lake County and the CTA's Orange Line to Midway Airport, for example, illustrates the regional commitment to bringing rail transit to new suburban and urban constituencies.

Several proposals for new rail service could substantially improve regional mobility.

Anticipating robust suburban growth, the Regional Transportation Authority is today shifting its emphasis away from existing services that have linked suburbs and downtown Chicago for generations. Precipitating its shift toward a more comprehensive focus was the release of the Future Agenda for Suburban Transportation proposal in 1992. As a result of this initiative, Metra is moving forward with a 10-mile extension of the Union Pacific West Line from Geneva to Elburn, one of the area's fastest growing communities. A circumferential commuter route along the Elgin, Joliet & Eastern Railway, which traverses Chicago's collar counties, is also receiving strong consideration, despite opposition from residents in South Barrington and other neighboring communities. Toward this end, the recent Chicago Area Transportation Study report *Destination 2020* (a mandated triennial report that assembles and lists transportation projects eligible for federal funding) calls for service to begin along a 50-mile stretch of the 103-mile route that would run in a semicircle around Chicago from Waukegan to Elgin, then to Joliet and through the south suburbs.

This expansive project would link with several Metra lines extending from Chicago while providing service to the rapidly increasing number of suburb-to-suburb travelers. Other current proposals include new routes to south and southeastern suburbs such as Manhattan, Steger, and South Holland, extending and improving the Metra Electric route to the proposed airport at Peotone, and providing service into central Kane County. After years of study, the metropolitan area appears to be moving toward a truly coordinated regional network.

Another promising development is the proposed high-speed rail system that would make Chicago the center of a 3,000-mile network throughout the Midwest. The federal government has authorized funding for 11 routes emanating from Chicago to such places as Detroit, St. Louis, Minneapolis, and Cincinnati (via Indianapolis). This expansive system, expected to be finished in 2006, could attract 8 million passengers by 2010, allowing expected revenue alone to offset the $3.5 billion in capital costs. State investments in roadbed and signal improvements will soon permit 110 mph operation over some sections of track to St. Louis and Detroit, vastly reducing travel times.

Safety considerations are serving to galvanize policymaker interest in railroad investment.

Any long-term expansion of our rail-passenger system must squarely address safety concerns, especially those related to grade crossings, where the majority of accidents occur. Due in part to the successful national public-education campaigns of Operation Lifesaver, crossing accidents dropped 21 percent between 1992 and 1997. Over this same period, fatalities fell more than 20 percent. Similarly, Metra's overall accident rate has declined 31 percent per million train-miles between 1992 and 1997. Nevertheless, notorious accidents, such as the heavily publicized Amtrak collision north of Kankakee in 1999, have heightened public interest in more aggressive government policies. Further progress will require continued education campaigns, improved warning systems, and increased enforcement of violations.

Technology ultimately may be the key to substantially improving rail safety. Federally sponsored technological innovations, including gate systems that reach across all traffic lanes and median barriers that prevent illegal automobile crossings, will soon be available. Metra has played an important role in this research, testing new motorist-warning systems, signal boxes, and camera-surveillance demonstration projects, and further contributing to ongoing pilot studies. Toward this end, Illinois legislated in 1996 increasing penalties for motorists who drive around lowered gates. With recently approved increases in funding for Operation Lifesaver and access to better technology, rail travel appears capable of progressively improving its safety performance over both the near- and long-term.

The cost of rebuilding and expanding rail infrastructure remains an unresolved regional financial issue.

Capital investment, cooperation with the private sector and better marketing are essential components in fostering a more responsive and efficient rail transit system. Currently, state officials have yet to allocate $5 billion of the Regional Transportation Authority's projected budget of $15 billion for the next 20 years. Although the Illinois FIRST initiative, launched by the state in 1999, will substantially boost transit funding, the outlook for many critical projects remains uncertain.

Informed decisions are necessary now to head off a shortfall. Several industry commentators have proposed shifting the true costs of driving to motorists through such measures as increased fuel taxes and registration fees. Another, less contentious option is to continue making incremental improvements, such as remodeling transit stations, providing additional rail-bus connections, and using the mass-media outlets to promote the advantages of choosing rail transit. As transit agencies experience fiscal stress, most agree that they should turn to competitive contracting to bolster efficiency and improve service quality. These ideas are just a sample of measures that can enhance the role our transit network plays in the region's economic future.

Chaddick's concerns have been addressed on many levels, but the magnitude of the problems facing transit officials suggests that this will remain a high-profile issue for years to come.

5

Adequate and Affordable Housing

Policies, Plans, and Action to Ensure Adequate, Affordable Housing

"Decent housing at a reasonable cost has become a priority issue in the Chicago area. The proportion of dwellings without plumbing, electricity, heat, and other basic amenities may have plummeted dramatically since the 1950s, but the major housing problem confronting both the poor and middle-class is now affordability—how much of their income they need to pay to keep a roof over their heads. Since 1980, incomes have not kept pace with housing costs, leading to a growing squeeze on renters and homeowners alike.

"A proposed housing plan or element should constitute a primary part of the comprehensive plans of the city of Chicago (and other municipalities), Cook County, and the state of Illinois, and these plans should attempt to coordinate and better utilize the many available federal and state housing assistance programs.

"We should be seeking definitive answers to the following questions:

- *What role should be played by the development controls (zoning, building code, subdivision regulations) in providing needed housing?*
- *How can we best utilize innovative housing forms such as factory-built housing, cluster housing, and zero lot line development?*
- *What are appropriate densities and dwelling types?*

"Any and all barriers to achieving affordable housing must be dismantled. Every Chicago resident deserves decent and safe housing at a reasonable cost."

— Harry F. Chaddick

Contemporary Assessment

The sense of urgency felt by policymakers to foster the construction of affordable, decent housing has profound implications for urban and suburban constituencies. Many officials and residents consider the provision of such housing one of the factors most crucial to the vitality of our region. Although a comprehensive solution to this problem has yet to be found, Chaddick would likely find Chicago's emerging policy measures both inspiring and more promising than past efforts—which ended in such institutional turmoil that the federal government intervened and ultimately assumed control of all Chicago Housing Authority properties.

The four points below exemplify the many difficult concerns associated with this issue.

Policy changes at the state and federal level raise new questions about the availability of affordable housing.

Recent studies suggest that federal and state revisions to welfare policy could exacerbate the already growing deficit in affordable rental housing. Even as the country experiences sustained economic prosperity and declining welfare enrollment, funding for low- and moderate-income housing developments continues to diminish throughout the region.

The prevailing shortage of affordable housing can be attributed to both the deterioration of the existing stock and the removal of remaining low-cost rental units from the marketplace. Adding to the problem, many landlords find maintenance costs almost prohibitive as older structures near the end of their useful lives. A 1997 Harvard University study found the rate in which affordable housing is being lost across the country is likely to accelerate in the years ahead. Many local analysts feel the same economic forces responsible for the revitalization of a great number of urban neighborhoods also are leading to dramatic increases in property taxes, thereby hastening the elimination of affordable housing units from the market. With estimates showing that more than 100,000 low-income renters in Chicago are unable to find affordable apartments, advocates have consistently called for greater funding for public housing agencies.

This problem is exemplified by a 1997 National Low Income Housing Coalition study which shows that fully half of all renters in our region are unable to pay average rents and are forced to give up other necessities to meet housing costs. The study also estimated that 50 percent of Chicagoans pay more than 30 percent of their incomes for rent—the standard the federal government uses to determine housing affordability. Such evidence poignantly shows that the average income of many renters has failed to keep pace with rising rental rates.

Although local housing initiatives have helped temper the loss of affordable housing units, the demolition of aging high-rise housing structures has notable long-range ramifications.

According to the Chicago Rehab Network, a coalition of housing organizations, nearly 114,000 predominantly rental units of affordable housing disappeared between 1970 and 1990. This number suggests the predicted national shortage has affected Chicago sooner than many other cities—largely due to recent federal housing policies calling for the demolition of many public housing buildings.

Recognizing the profound implications of this problem, officials in 1994 undertook a five-year $750 million plan to develop or preserve 40,000 housing units for low- and moderate-income residents. As this program gathered momentum, however, the mandated federal demolition of antiquated high-rise units progressed. In 1994 and 1995, for example, the city sponsored the construction of 5,200 units while tearing down 7,000 over the same period. Although the demolition has continued, the initial five-year plan exceeded its goals and has been lauded even by housing advocates who are frequently critical of city policy. Such an outcome bodes well for proposals to further reduce the city's reliance on deteriorating high-rise units in the years ahead.

The 1999–2003 affordable housing plan calls for a budget increase to $1.3 billion and for expansion in several areas of the previous measure. To develop a broader understanding of this issue, the city invited input from more than 30 housing advocates. Recognizing the need to work cooperatively with developers, the city is also offering additional incentives for those who build three or more bedroom units for families with incomes less than half the neighborhood median. (Of the 1,063 new rental units the city subsidized in 1998, only 408, or 38 percent, were suitable for families.) In addition to concentrating on family units, the plan intends to produce higher quality housing stock.

Officials anticipate that this shift toward producing larger, higher quality units along with increasing construction costs will result in the production of 10,000 fewer units than the first plan. The city hopes to use its resources more selectively to focus on neighborhoods and residents in the greatest need.

An infusion of both public and private sector capital is laying the groundwork for successful housing redevelopment.

City officials and private agencies continue to forge bonds and seek out creative solutions to many ongoing housing issues.

Community development corporations serve as the nucleus of many of these institutional partnerships. Such organizations assist in the revitalization of low- and moderate-income neigh-

Deteriorating conditions blight Chicago Housing Authority property on Chicago's West Side.

borhoods, serve as conduits for innovation, and according to officials, are involved in more than 50 percent of the projects handled by the Department of Housing. Community development corporations help leverage funding for development with intermediaries such as the Local Initiative Support Corporation. They also coordinate corporate investment through the Low-Income Housing Tax Credit—a principal source of support nationally that gives corporations a tax break for investing in development projects. Another promising collaborative tool, the City Mortgage program, helps low- and moderate-income families buy homes with lower-than-market-rate mortgages. These examples suggest that innovative partnerships can bring grassroots constituencies into the decision-making process, thus ensuring that public investments are responsive to neighborhood needs.

The experiences of recent years suggest that the most effective methods to improve housing conditions involve a mix of both supply- and demand-side strategies.

Policy analysts throughout the Chicago region recognize that the changing characteristics of the market will necessitate fundamental changes in their approach to providing affordable housing.

The *supply-side* measures focus on rehabilitating the subsidized housing model that served to isolate low-income families. They are based on the premise that units of government can promote the integration of its diverse residents by replacing high-rise development and encouraging public- and private-sector partnerships to create and oversee mixed-income developments. Conversely, *demand-side* strategies focus on expanding housing choice by providing vouchers that allow residents the freedom to select where they wish to live. They enhance the customer's ability to pay for various types of housing stock.

Municipalities seeking to provide affordable housing often face troublesome issues surrounding the application of land use and building controls. If a municipality's regulations are developed independently from a comprehensive housing plan, they may contain provisions that require unnecessary materials, design features, or large lot sizes, thus discouraging the construction of more affordable types of housing. Many suburban communities are sensitive to the fact that development controls can directly affect the supply of affordable housing, either by facilitating opportunities to increase its supply or by deterring its development through exclusionary restrictions. Nevertheless, building codes and other forms of development control continue to drive up the cost of living in many suburban areas.

Residents of our region who are eligible for subsidized housing and their advocates can find encouragement in the direction set forth by the Commercial Club of Chicago's 1998 *Metropolis 2020* report, which recommends using a combination of the above strategies. The plan calls for the expansion of the Section 8 voucher and relocation assistance programs in conjunction with the demolition and replacement of inappropriate or decaying low- and moderate-income housing stock. While barriers such as exclusionary zoning, insufficient counseling of voucher program participants, and biases against a perceived threat to property values endure, governments at all levels recognize the need for creative market-driven solutions to increase access to affordable housing.

In summary, Chicago has made affordable housing a priority in both word and action, mitigating a crisis that seemed inevitable only a few years ago. Although the ultimate goal of adequate housing for all remains elusive, Chaddick would consider recent endeavors an important stepping stone to more significant policy initiatives in the years to come.

6

High-Intensity Land-Uses Plan

Master Plan for High-Intensity Major Developments or Activity Centers

"Chicago is, perhaps, the most vibrant and constantly changing major city in North America. It seems that the Greater Chicago community is continuously reshaping itself, planning for and attracting major new high-intensity activities that usually benefit all of us when completed, but often put us 'through the wringer' in the process.

"There is a critical need in the metro area for some type of master plan for proposed major activity centers because these giant developments generate so much traffic, jobs, and economic growth. Both positive and negative effects merit careful planning. This master plan for high-intensity activities could well become a component of the region's comprehensive plan and afford everyone the opportunity to study the various proposals and possible effects and voice their comments and suggestions.

"Choosing the site for a third airport or the location of major sports stadiums will certainly have a significant impact on the entire region, as will the timing or phasing of such developments. Even the siting and design of regional shopping centers and industrial parks or entertainment complexes can have far-reaching and dramatic impacts upon surrounding land-use.

"A different form of major development or redevelopment that warrants constant scouting and 'planning' because of the overwhelming role it plays in determining our lifestyle and enjoyment is, of course, our lakefront and our river frontages. We should possess detailed,

well-thought-out plans for the use and maintenance of our waterways and land abutting them."

— Harry F. Chaddick

Contemporary Assessment

Large-scale development projects often affect the very essence of neighborhoods, abruptly changing their appearance and character. As Chaddick noted, a largely piecemeal planning approach has, to date, prevented policymakers from addressing the full spectrum of issues accompanying projects with the capacity to alter the rhythm of community life. Without a general plan for the region, officials are often forced to consider projects without the holistic perspective necessary to promote harmonious land-use and efficient traffic flow.

Throughout the early 1990s, numerous planning agencies espoused the concept of using diversified regional centers, that is, places with relatively high-density development that serve as hubs for transportation, communication, and other forms of infrastructure. Many called for encouraging major developments to establish themselves within or near diversified regional centers as a means of fostering the efficient allocation of public resources. Unfortunately, the public policy debate revolving around high-intensity suburban developments has tended to move away from the concept of diversified regional centers in recent years.

The need for more thoughtful analysis of the performance of high-intensity land-uses is exemplified by the following examples, which describe four major urban planning projects: the River East development project, State Street's redevelopment, efforts to improve the lakefront, and the proposed airport at Peotone.

The River East project illustrates why fiscal incentives for promoting development, especially large-scale projects, must be accompanied by policies to protect the quality of life that neighborhood residents have come to expect.

This Near North neighborhood demonstrates the extent to which optimism about market opportunities can obscure the formidable risks associated with a major project. Although the outcome of the project remains to be seen, the nature in which Chicago evaluated and approved it suggests a need for a more comprehensive planning process. In the enthusiasm of the moment, the city evidently chose to initially surrender the role of methodically evaluating the project's long-range effects, which could have left the city and its residents overextended in the future.

This enormous 10-million-square-foot mixed-use development, originally slated to cover 13 acres between North Michigan

Chicago's rapidly evolving Streeterville neighborhood is the site of high-intensity River East development.

Avenue and Navy Pier, is part of a broader rehabilitation effort at Cityfront Center. The city approved a planned unit development project in the area in 1985, well before the onset of the current construction boom and the transformation of Navy Pier into Chicago's most popular tourist attraction—drawing more than 8 million visitors annually. Density of development was of little concern when the city originally agreed to zone the area for up to 6,000 residents.

The unveiling of the 10-year, $1 billion River East project's design in 1997, however, raised vociferous protests from area residents. Many expressed apprehension about the impending "Manhattanization" of an important downtown tract. Officials and neighborhood groups expressed serious concerns about how the project would affect the lakefront and riverfront and its implications for the quality of life. These concerns revolved around the fact that area streets would be surrounded by formidable buildings lacking both a distinctive character and convenient transit access. Hoping to prevent an addition to the aesthetically barren "canyons" that dominate many of the commercial areas of Chicago, many commentators and residents called for the amendment of the original planned unit development agreement.

With the River East developer reportedly needing only a building permit, the city began exploring traffic, parking, waterfront access, and other quality of life concerns. The city and developer reached an agreement after a year of negotiations, which included a controversial legislative proposal to downzone the entire area. This move would have effectively killed the entire project due to increased restrictions on permitted land-uses and densities for the site. The eventual compromise on a scaled-down version of the project to 7 million square feet included increases in open space, taller and slimmer buildings that use less land, fewer total units, larger setbacks to retain better views, and transportation improvements.

In the end, by evaluating longer-term issues of density and traffic congestion, the city succeeded in reducing the project's magnitude and compelled area developers to join a transportation management association to formulate a comprehensive transportation plan for the thriving area.

The River East agreement required the city to effectively critique a large project with far-reaching implications under the pressure of approaching deadlines and possible legal action. The city did not have the benefit of a clear set of planning guidelines that could have eliminated significant duplication in the decision-making process. The current process can deny the developer of the near-term flexibility necessary to be successful and limit opportunities for the city to refine and promote appropriate projects.

The redevelopment of State Street demonstrates how the central business district must evolve with the times to remain viable as a high-density retail area.

Few cities across the nation boast of a commercial corridor as historic and diverse as Chicago's State Street. After relinquishing its status as the nexus of downtown commercial life in the 1970s and 1980s, State Street offers valuable lessons for planning officials searching for ways to reinterpret the community's assets.

Jane Jacobs asserts in her classic *The Death and Life of Great American Cities* that a healthy city is an amalgamation of diverse and ever-changing land-uses. In order to achieve a beneficial variety of land-uses, an area must be flexible enough to support multiple "primary" functions, such as office space and retail, while allowing secondary functions, such as restaurants and cultural facilities, to emerge in support of these principal functions.

Marshall Field & Company's flagship store on State Street displays its holiday decorations. *(Credit: Peter J. Schulz/City of Chicago)*

For many years, State Street exemplified this diversity, accommodating a wide array of retail and business firms. In the 1970s, however, State Street emerged as a symbol of the vulnerability of traditional downtown districts to economic change—an issue of grave concern to Chaddick, who espoused bold new approaches to fostering retail development. Today, as leaders direct their efforts toward restoring retail activity in the urban center, State Street is poised to recapture some of the luster of its storied past.

With an eye to the thoroughfare's proud commercial heritage, planners explored ways to integrate new land-uses with existing structures. Working closely with area business groups, officials produced a comprehensive plan, *Vision for Greater State Street: Next Steps*, that met their needs for a site-specific set of development standards and aesthetic guidelines. This plan established standards for incorporating new construction with the area's historic character and synthesized a broad action plan for strategically filling vacant sites, thereby creating a smoother transition from the retail-oriented section at the southern end of State Street and an entertainment-oriented area to the north.

This coalition of public and private sector stakeholders in the area facilitated State Street's return to prominence. Having successfully recovered from the failed attempt at being remade as a suburban-type pedestrian mall, this area is once again a vital component in the recent resurgence of retail activity in the central city. Without a regularly updated downtown plan to coordinate similar efforts throughout the area, however, these success stories run the risk of occurring in a planning vacuum. For evidence of the potential problems, planners need only to examine the lack of convenient transportation links between the State Street and North Michigan Avenue retail areas—a problem partially attributable to the demise of a planned light-rail "circulator" system slated to serve the downtown area.

The Museum Campus provides a popular vantage point to view the downtown skyline. *(Credit: Peter J. Schulz/City of Chicago)*

The Chicago lakefront exemplifies how the absence of comprehensive planning can compromise long-term goals.

The decision-making process regarding public land along Lake Michigan has been relatively disjointed over the past several decades—with outcomes that often threaten to compromise larger municipal goals.

With 30 miles of coastline stretching from Evanston to Indiana, Chicago's lakefront is a treasured community resource. It has suffered in recent history due to the lack of a unified plan governing its development and maintenance. Although the 1972 Lakefront Plan established a clear precedent for planning along the shoreline, it is now woefully outdated. Nevertheless, it remains the only coordinating tool providing oversight for development decisions along the entire shoreline.

Although the Park District, working closely with a community group, crafted a plan for Lincoln Park in 1995 and commissioned a 1998 effort to write a comprehensive plan for the South Shore's Burnham Park area from Roosevelt Road to 57th Street, there remains no larger vision statement binding the larger goals of the entire lakefront. The closest approximation to such an endeavor would be the open space plan the Park District and the Forest Preserve District published in 1998. This expansive document addressed many park and waterway issues, but provided only a broad list of objectives concerning the lakefront.

Blair Kamin's Pulitzer Prize-winning essay series, "Reinventing the Lakefront," published in the *Chicago Tribune* over several months in 1998 and 1999, weighed in on topics ranging from the underutilization of Grant Park to the inaccessibility of the shoreline to workers in the Loop. This illustrated series drew considerable public attention to the social and economic consequences of these missed opportunities, which often leave our parks nearly bereft of activity.

Although more than $500 million is earmarked for lakefront improvements over the next decade, a fragmented decision-making process could limit the success of the various projects under consideration. Several state legislators and commentators have suggested the current mayoral administration develop a commission to foster coordination between agencies with a stake in the future of the lakefront. These same organizations demonstrated the value of such coordination when they joined forces to facilitate the rerouting of Lake Shore Drive and the creation of the Museum Campus, a process under the stewardship of a similar type of oversight commission. The enormous benefits of this collaborative effort, along with the high expectations accompanying several pending large-scale projects—including the redesign of Grant Park, the opening of Millennium Park, and the approval of a dedicated busway along the lakefront from Randolph Street to McCormick

Place—illuminate the need to create a broad strategy to coordinate developments with far-reaching civic implications.

The debate surrounding the proposed airport at Peotone demonstrates how a lack of consensus about the benefits and costs of high-intensity development on the urban fringe can prevent timely decision-making.

The contentious and divisive debate surrounding proposals to establish a new international airport at the southern periphery of the metropolitan area near Peotone is emblematic of our region's growing inability to evaluate large-scale infrastructure projects in a timely fashion. As the debate becomes more polarized—and land envisioned for the airport is gradually subsumed by commercial and residential development—environmentalists and commercial groups have been able to find little common ground.

Both groups continue to disagree about virtually every aspect of the airport proposal, from its costs and commercial viability to its effect on the decentralization of the region's population. Discussion among stakeholders also has served to magnify the level of mistrust between the city of Chicago and the surrounding suburbs. The issue's complexity is increased further by seemingly irreconcilable differences of opinion about the problem of aircraft noise adjacent to O'Hare Airport—a problem enshrouded in regional politics that severely limits options for addressing the need for improved mobility.

It is impossible, of course, to do more than summarize the debate revolving around the need for a third major airport in this volume. However, readers should understand that the airport's proponents are moving away from their original plan for a two-runway facility. This plan was originally denied federal funding, having been left off the national airport planning list in 1997. In response, state officials recently submitted revised plans for a smaller Peotone airport with a single runway and 12 gates on 4,100 acres, which is predicted to receive 1 million passengers in its first year of service. However, even this scaled-down proposal was coolly received by federal authorities, garnering little support from both private investors and airline carriers operating out of O'Hare and Midway airports.

Supporters of the new airport assert that Peotone would strengthen the position of the other airports by allowing our region's air service to expand in response to rising demand. It would ensure that the region will be able to accommodate predicted increases in air traffic, allowing O'Hare to foster its standing as a major international gateway, while Midway and Peotone meet growing demand for domestic flights. Suburban communities as well as various councils of government, who are among the project's most vocal proponents, also herald the project's potentially beneficial effects on the rate of economic growth in outlying areas. Nevertheless, opponents are adamant in their view that the south suburban airport will diminish the economic vitality of Midway and O'Hare—both of which are undergoing renovation to meet forecasted growth. Others assert that plans for Midwestern high-speed rail, as well as anticipated changes in air traffic control which could dramatically reduce the separation of takeoffs and landings at existing airports, have the potential to render the new facility unnecessary at least in the short-run.

Our region's often cumbersome approach to decision-making virtually assures that the debate about additional airport capacity will remain volatile and laden with emotion. When establishing policy positions, suburban entities have little incentive to take into consideration the full extent of the probable diversion of economic activity from established commercial areas—or the benefits that might flow to communities outside their region. At the same time, despite forecasts for air travel growth (projected by the Federal Aviation Administration at nearly 4 percent annually) that suggest a pressing need for additional aircraft capacity, Chicago officials will almost certainly remain resolved in opposition to the airport, largely on account of its potentially damaging effects to the city's revenue base.

Recent developments appear to be giving Chicago the upper hand in the airport-planning debate. Through the auspices of the Chicago/Gary Regional Airport Authority, it is helping elevate the stature of the Gary/Chicago Regional Airport, another facility touted as a potential regional air transport facility. In late 1999, Pan American Airways launched jet service to several eastern cities from this airport, located 40 miles south of the Loop. This move bolsters the claim that this site could help ease the traffic burden facing O'Hare and Midway. Investments by area governments in promoting the airport as well as the construction of a new access road raise further optimism about the expansion of airside operations along Lake Michigan's southern shores.

In a wider sense, the debate surrounding Peotone illustrates the importance of establishing a broad statutory foundation for regional decision-making, especially relating to projects involving high-intensity land-uses. The unwillingness of some municipal governments to support regional planning goals is the source of both inefficiency and inequity in many areas of public policy, such as the creation of industrial corridors, expressways, and transit lines.

The Chicago region deserves a great deal of credit for its efforts to see major development projects through to completion, such as Deep Tunnel, the reconstruction of expressways, and new terminals at O'Hare. However, River East, State Street, the lakefront, and Peotone illustrate some of the challenges facing planners as they attempt to integrate large-scale development projects into the fabric of urban and suburban life. All portray the potential advantages of a binding comprehensive plan that sets out basic guidelines to determine the value and overall impact of such projects.

7

Planning Statutes and Development Controls

Examination and Revision of Illinois Statutes and Regulations Related to Planning, Development, and Controls

"In 1979, the Illinois legislature commissioned a compilation of state laws relating to development and planning. The result was a 1980 document entitled Illinois Laws Relating to Planning and Development; *a laudable expenditure of funds because it made it painfully clear that Illinois is operating under a terrible hodgepodge of confusing and fragmented pieces of legislation pertaining to municipal, county, and township powers to plan, zone, and plat.*

"Since the publication of the compilation, there has been no real effort to examine the effectiveness of the laws or to revise them in any way. A study committee was formed in 1985 to analyze the situation and recommended modification but was disbanded after a short period. This sorry condition is further exacerbated by the fact that the state of Illinois does not possess a land-use plan, nor any clear system of monitoring land-use and planning on a statewide basis.

"We would benefit greatly from some form of Illinois state land-use and development plan (comprehensive or concept) coordinated with a modern and efficient network of enabling acts and related legislation dealing with the planning and zoning powers and responsibilities of local government."

— Harry F. Chaddick

Contemporary Assessment

Understanding that change is a beneficial (if often difficult) part of civic life, Chaddick advocated revising state land-use statutes regularly in response to the evolving urban environment. He would certainly continue to be a vocal critic of the fragmented nature of land-use oversight in our state today. According to many observers, this fragmentation is accentuating the problems associated with the rapid decentralization of the regional population. Rising public concern about the social and ecological consequences of this process is enhancing policymaker interest in strengthening state legislation to facilitate more effective regional planning.

The following paragraphs offer a brief summary of the status of land-use governance in Illinois.

Current planning statutes are, to a large degree, unsupportive of the need for decision-making at a regional level.

The land-use enabling legislation for most states, including Illinois, is rooted in the 1920s publication of the *Standard City Planning and Zoning Enabling Acts.* As U.S. cities were experiencing unprecedented growth during that period, states widely adopted these acts in an effort to cope with rising population and urbanization as well as increasingly concentrated land-uses. The appropriateness of some of this legislation, however, diminished after World War II, when suburbanization and rising automobile use fundamentally changed the regional landscape. Nevertheless, much of the legislation was left either unchanged or altered only incrementally.

The current system gives Illinois municipalities nearly absolute control over land within their borders as well as certain "extraterritorial" authority over land up to one and one-half miles outside their boundaries. While county planning departments and

A residential neighborhood in Hoffman Estates illustrates a conventional development pattern under prevailing land-use law. *(Credit: Charles Hanlon)*

regional councils of government may have political mandates to promote harmonious development, the authority for most forms of development control ultimately rests with municipal officials.

At the same time, it is important to note that the existing system has certain advantages over a more centralized system. For example, it provides a degree of competition between municipalities in the economic development process that fosters greater responsiveness to the needs of citizens and institutions. By keeping decision-making at a local level, it encourages municipalities to tailor local services to the specific preferences of the citizenry. Indeed, in microeconomic theory, this is the basis of the Tiebout Hypothesis, which holds that communities tend to specialize by providing particular bundles of public goods. As suburban governments pursue economic development opportunities and politicians within them vie for voter support, units of government have incentives to be responsive to the demands of local constituencies.

More research is needed before one can make definitive conclusions about the consequences of existing municipal growth-management policies. Readers interested in exploring the broader microeconomic dimensions of various forms of development control are encouraged to consult studies on land-use control cited in the reference section, especially works by Altshuler and Gomez-Ibanez (1993), Brueckner (1999), and Fischel (1985).

As traffic congestion worsens and municipalities engage in intense competition over possible sources of tax revenue, many have called for new legislation to strengthen the role of county or regional planning agencies.

Interest in fostering sustainable development has been the impetus for significant legislative reform measures.

Beginning in the 1970s, several states examined and eventually revised, at least partially, their planning statutes to promote more environmentally sensitive forms of development. Their efforts are part of a national movement to modernize approaches to land-use governance. Illinois contributed to this movement with the 1976 publication of *Illinois Laws Relating to Planning and Development.* The era's reform fervor, however, largely dissipated by the 1980s.

Several current initiatives in our region also merit acknowledgment. The Northeastern Illinois Planning Commission has led a three-year project to create a regional growth strategy. This effort is intended to serve as a development timeline to help promote sustainable expansion over the next two decades. The Commercial Club and the Metropolitan Planning Council are prominent among those working to build coalitions and champion the benefits of environmentally responsible forms of development. Notably, the state legislature recently appointed a "smart growth" task force that has published a report describing the alarming loss of agricultural land and voiced support for the creation of a state planning office.

These efforts are taking place in concert with the federally supported Growing Smart℠ initiative of the American Planning Association. Since the inception of the initiative in 1994, planning professionals have worked to evaluate the statutory tools presently available to state governments and drafted model legislation that can be tailored to each state's needs. The American Planning Association also published in 1996 a Planning Advisory Service report that presents a series of research papers exploring the difficulties commonly associated with the existing planning legislation.

Chaddick would likely support the ongoing national efforts of the American Planning Association as well as the regional dialogues that seek mutually beneficial approaches to the management of land.

City of Chicago Needs

8

Comprehensive Planning

A Truly Comprehensive Plan for Chicago

"Beginning in the late 1950s, the city of Chicago has periodically issued 'land-use plans' for the city—usually prompted by an effort to justify major zoning overhauls or massive development proposals. Prominent among these plans are the collection of 16 Development Area Plans prepared in 1966–68 and the Chicago 2000 Plan *issued in 1980. Although these were well-intentioned efforts and contained valuable insights and projections, they amount to watered-down versions of a truly comprehensive plan.*

"Chicago needs and deserves a 'plan' that is comprehensive both geographically and in the scope of technical content. Future land-use is of course crucial, but of equal importance are other plan components such as transportation, recreation, public utilities and facilities, housing, education, and economic development.

"The optimum plan should be composed of both a short-term element (action plan: one to five years) and the more traditional long-range plan looking 20 to 25 years to the future. Any plan we consider must involve not only local officials and staff; we have to generate the meaningful participation of neighborhood organizations and residents as well as the business and industrial community, and this should be done early in the program.

"Whatever form the plan takes, it is essential that it be kept current and valid through periodic review and updating. All too many municipal plans are adopted and then reside forever (dusty and untouched by human hands) on some commissioner's bookshelf."

— Harry F. Chaddick

Contemporary Assessment

Contemporary planners in Chicago are in many respects protégés of Daniel Burnham and Edward Bennett, whose 1909 *Plan of Chicago* pushed both the city and the planning profession into an entirely new realm. As the above passage suggests, Chaddick was among those promoting comprehensive planning efforts that held the promise of bringing Burnham's ideas to fruition. The legacy of comprehensive planning in Chicago, however, is understood by today's civic leaders to be somewhat bittersweet. While it has given rise to much ambitious thinking, municipal planning remains dominated by a patchwork of smaller plans, many of which sorely need revision and modernization.

The following paragraphs briefly review several important developments in this continuing dialogue.

Though nearly a century old, Burnham's plan continues to inspire planners and provide a frame of reference by which officials and private groups evaluate plans.

Burnham's ideas ushered in a bold new era of urban planning that had profound implications for successive generations. His visionary design brought to Chicago, for example, a system of boulevards that linked parks throughout the city and set aside vast tracts of land for public use along the lakefront. Over the ensuing decades, many planning officials made bold attempts to integrate concepts from Burnham's original plan into their efforts, thus promoting continuity in the urban development process.

Among the many notable additions to the city's large-scale planning efforts, the 1966 *Comprehensive Plan of Chicago* contained Mayor Richard J. Daley's sweeping vision for the future. This plan vividly described proposals to create 10 self-contained neighborhoods around the city that would support growth in the retail and manufacturing sectors. However, such an outcome was never realized. The tumultuous social events of the period, especially those revolving around the civil rights movement, tended to relegate civic planning to the margins of municipal affairs. Coincidentally, the Gautreaux decision, which halted any further construction of public housing structures, also occurred in 1966, the same year the plan was released.

The effort to promote comprehensive planning reached another milestone in 1973 with the release of the *Chicago 21* plan. This expansive document, published in the midst of the rapid suburbanization of the region, constituted the first formal plan focusing primarily on the 11 square-mile central area of the city. Intended to expound on the 1966 plan, this compendium boldly confronted such difficult issues as historic preservation and the need to enhance Chicago's cultural mosaic at the neighborhood level. Furthermore, it introduced the concept of the central area as a potential home to a vibrant residential population where commercial and business interests had long dominated. A significant legacy of the *Chicago 21* plan is Dearborn Park. This successful residential development in the South Loop transformed blighted industrial property occupied largely by old railroad yards once serving Dearborn Station.

The most recent comprehensive plan for this area was the *Central Area Plan,* published in 1984 by the Central Area Committee. This plan illustrated the flurry of civic interest in establishing Chicago as the host of the 1992 World's Fair. Providing an updated timeline for many projects called for in the 1973 plan, this effort incorporated the burgeoning service and technology sectors and initiated the development of the Museum Campus and the relocation of Lake Shore Drive.

Each of these comprehensive plans contributed to ideas that helped form the basis of contemporary city policy. However,

Chicago's Museum Campus, as seen from Lake Michigan, is home to the Field Museum, Shedd Aquarium, and Adler Planetarium. *(Credit: Peter J. Schulz/ City of Chicago)*

like all plans, circumstances and changing times eventually rendered much of their content obsolete. This situation has left officials with the burdensome task of considering development proposals on a largely case-by-case basis. To some degree, this approach is necessary given the myriad details associated with many of the city's extensive projects. Nevertheless, many commentators rightfully feel an overall vision is necessary to bind together these disparate projects into a coherent agenda that best serves the community.

While the promise of creating a new comprehensive plan remains unfulfilled, recent efforts are moving Chicago indirectly toward such an outcome.

The current mayoral administration has worked closely with neighborhood groups and public agencies to implement several area-specific planning efforts. These plans, involving such high-profile districts as Maxwell Street, River East, and the Near West Side, have generally succeeded in responding to the particular needs of individual neighborhoods. A noteworthy consequence of these efforts is heightened civic interest in developing for the entire city yet another comprehensive plan—one capable of addressing planning issues well into the new millennium.

Fueling optimism about such comprehensive planning measures is growing sentiment toward "regionalism," an approach to planning issues that focuses on improved cooperation among the area's municipalities. This movement draws its strength from public concern about the increasingly decentralized pattern of municipal development throughout the Chicago region. Proponents of regionalism buttress their arguments by citing estimates from the Northeastern Illinois Planning Commission showing that, between 1970 and 1990, the six-county region's population grew by only 4 percent while the acres of developed land increased by 33 percent during the same period. This finding, perhaps more than any other, has set the tone for the debate about land consumption in this region.

Contributing to rising public concern is the apparent inefficiency of the existing—and often fragmented—system of governance throughout the region. Among the difficulties that many consider inherent under the system are intractable problems relating to traffic congestion, a shortfall of housing for the moderate- and low-income population, and a growing separation between areas with affordable housing and major job centers for less-skilled workers.

The resonance of these concepts among area business and civic leaders is evident in *Metropolis 2020,* a 1998 publication of the Commercial Club of Chicago (the same organization that commissioned Burnham's work). This two-year project draws on the expertise of many local organizations and business leaders, and provides a framework for fostering economic and social development throughout the region in coming years. Another significant initiative, Regional Growth Strategy from the Northeastern Illinois Planning Commission, seeks to improve the region's quality of life and promote sustainable forms of development.

Such measures build upon the tireless efforts of many organizations to bring land-use issues to the forefront of the civic debate. Prior to this recent initiative, the Northeastern Illinois Planning Commission, for example, published in 1992 the *Strategic Plan for Land Resource Management,* a document articulating the need for greater coordination of development throughout the region. Also in 1992, Openlands Project and the Northeastern Illinois Planning Commission jointly published the *Northeast Illinois Regional Greenways Plan* that called for new greenways and connecting trailways throughout northeastern Illinois. Another influential report, *Destination 2020,* addresses transportation issues within the context of the metropolitan area's expansion. Published in 1998 by the Chicago Area Transportation Study, this document outlines a long-term strategy for addressing transportation problems in the six-county area. Equally noteworthy is the Campaign for Sensible Growth, a coalition headed by the Metropolitan Planning Council. The Campaign for Sensible Growth espouses more efficient and environmentally responsible development, articulating its position in its 1998 publication *Growing Sensibly: A Guidebook of Best Development Practices in the Chicago Region.* At the municipal level, the recently convened Metropolitan Area Mayors' Caucus is promoting innovative approaches to solving problems affecting Chicago and surrounding suburbs.

These are just a portion of many initiatives for which Chaddick would undoubtedly be a vocal proponent. Only time will tell whether they substantially change the course of real estate development and land-use control in the metropolitan Chicago area—or set in motion new legislation that encourages collaboration between governments in their role as stewards of the land. Taken as a whole, however, these efforts apparently signify a remarkable paradigm shift in regional planning—one broadly consistent with the venerable 1909 plan created by Burnham and Bennett. They hold the promise of perpetuating the region's reputation as a flourishing and distinctively American metropolis.

9

Updated Land-Use Survey

Current and Detailed Land-Use and Environmental Survey

"Before Chicago can begin to shape meaningful plans for future development and growth, and prepare zoning and other codes designed to put those plans into effect, the city must have the necessary basic information and material to work with.

"An important first step would be a field survey of each building and lot in the city to record exact land-use, number of dwelling units, and for commercial and industrial establishments: product or service, number of employees, and lot and floor area. Those environmental conditions warranting inventory and evaluation include needed street, alley, and sidewalk repair; areas lacking adequate storm drainage; absence of curbs and gutters; poor property maintenance; pollution problems (noise, odor, smoke, visual); overcrowded areas; and ecologically sensitive sites.

"Is Chicago's existing information base for planning and development control up-to-date and accurate? The most recent comprehensive parcel-by-parcel land-use inventory was conducted in 1969.

"A comprehensive inventory of current land-use and conditions can be completed by trained college or high school students from each of our neighborhoods supervised by staff from the Department of Planning and Development. It would constitute an excellent community-involvement project and should be achievable within one summer."

— Harry F. Chaddick

Contemporary Assessment

Chaddick hardly could have imagined the degree to which our nation's shift to an information-based economy would affect the very essence of professional and personal life. In only a few years, the methods planners use to acquire and apply related data has changed immeasurably. Many aspects of geospatial data (i.e., data organized with respect to its geographic characteristics) that would have been available only in an exhaustive, paper-based survey have been largely incorporated into sophisticated and increasingly accessible geographic information systems (GIS).

Although Chicago and agencies interested in development issues still conduct local and issue-specific surveys, such as those used in corridor planning or tracking brownfields and vacant lots, GIS technology provides the opportunity to consider planning issues as part of the much wider system of land-use management and infrastructure oversight.

The following paragraphs describe briefly some of the more salient developments affecting the manner in which planning officials collect and analyze data about land-use.

Emerging GIS technology is giving rise to a veritable revolution in geographic analysis by the urban planning community.

The current boom in GIS technology is rooted in the resource-intensive aerial photographing studies conducted during the 1960s. Recent technological progress in the storage and application of data that can be spatially referenced is making vast data sets accessible to the general public. In previous eras, only the larger cities and units of government could justify the expense of creating comprehensive maps for planning, projecting, and

An aerial view shows downtown Chicago looking north from the South Loop district. *(Credit: Peter J. Schulz/City of Chicago)*

monitoring land-use and infrastructure development; today, nearly all communities can afford such systems.

Equally significant, high-altitude imaging has gained widespread acceptance in the development community over the past decade. This technology provides an effective tool to track important infrastructure components such as roadway usage, utility location, real estate platting, and many others. Maps of different factors can be cost-effectively created as overlays to evaluate interactions between various land-uses.

Contemporary GIS technology has enormous advantages over the manual systems that were pervasive as recently as the early 1990s. It allows users to digitally represent maps, photos, charts, or any other data; add different statistical or characteristic information about an area; and then generate a wide variety of spatial analyses. It permits communities to maintain their own database of spatial information and integrate it with other data-based references, such as tax assessments, crime incidence, or even health factors. Communities employing this technology can also provide public access to their systems, thus facilitating administrative processes such as remodeling permits or zoning variances. GIS also can help businesses choose appropriate locations or vividly illustrate local market dynamics.

While some municipalities find the investment required to develop a GIS and input all necessary data to be prohibitive, the financial barriers are gradually falling as computer hardware and software decline in price.

Chicago has dramatically altered how it manages land-use information.

Like all large cities, Chicago oversees an enormous amount of data about individual and group parcels of property. The challenge facing officials is to efficiently update and present this information in a meaningful way.

Overcoming such challenges can lead to tremendous gains for the community. In 1992, for example, officials commissioned the Chicago Citywide Infrastructure Management System study to address concerns about aging infrastructure—an especially poignant initiative following the Great Chicago Flood that year. The four-year, $10 million project created a GIS for Chicago's more than 13,000 miles of roadways, and water and sewage lines, helping to prioritize capital projects and identify possible problem areas. This effort and other ambitious projects established a link between modern desktop computer technology and a wide range of planning issues.

Attempting to raise its dated information management system to contemporary standards, Chicago in 1997 began the shift away from mainframe computers to a more decentralized, but fully networked, system. The city outsourced much of the work for the $17.5 million project, known as Citywide Database, that coordinates information for all 42 municipal operating departments. As part of this larger effort, it created a new database to store all GIS and mapping information for the city. This update should streamline many diverse planning functions for the Department of Planning and Development as it works in conjunction with other offices dealing with land-use issues.

The information technology available for data collection, however, has not eliminated the need for a more traditional survey of land-uses that considers details about the status of property within the city.

Despite the many improvements to its databases in recent years, Chicago still lacks an up-to-date inventory describing the microenvironment of various parcels of land, such as the type of business (e.g., grocery, dry cleaning, pharmacy) or residence (e.g., single-family, apartment) on each parcel. Collecting this information will require considerable administrative effort, as this data cannot be collected using aerial imaging or traditional zoning maps. It is important, however, that the city invest in such initiatives in accordance with the Standard Industrial Codes (SIC). The last survey that gathered such information dates back to 1970 and is now virtually obsolete.

Maintaining an accessible database that indicates the SIC for every lot would allow for more informed decisions regarding zoning amendments, urban development plans, and the analysis of neighborhood needs. Unfortunately, officials who lack such information must engage in the laborious practice of conducting "windshield surveys" to acquire specific information about the nature of business and industry within various zoning districts.

While strongly advocating such a traditional survey, Chaddick would likely also endorse the new mapping information technology that holds the promise of elevating the planning profession to an entirely new level. Although the times and technology have changed, Chaddick's essential assertion—that planners need current and detailed information to understand the value of land—remains beyond dispute.

10

Parking Needs

Comprehensive Study of Parking Needs In and Around Chicago's Central Area

"Through the years there have been numerous attempts to solve the parking problems of Chicago's central business district. These have included revenue-bond financed, privately operated parking garages—almost all of which have since closed—and the encouragement of motorists to use parking lots on the Loop's periphery combined with bus transportation to work locations. Thus far, none of these 'solutions' has proven satisfactory, thereby deterring downtown shopping and other business activities.

"People are willing to drive downtown to shop and complete business transactions if free (or nominally charged) parking is made available, as it is in the outlying shopping areas. We must examine all of our options for providing necessary parking, including such innovative possibilities as utilizing the downtown alleys and air rights.

"A complete inventory and study of the present parking situation is long overdue, and we must prepare an effective action plan to provide adequate parking before our core area develops terminal gridlock."

— Harry F. Chaddick

Contemporary Assessment

Chaddick would likely take little consolation in the progress that has been made on this issue. While studies suggest a need for more attention to parking, the number of available downtown spaces has remained relatively static in recent years.

Despite many years of vocal pleas from politicians, business leaders, and the general public for more accessible and affordable parking, the situation has only recently begun to improve.

The paragraphs below touch on several important developments related to the availability and cost of parking.

Both policy and marketplace developments are responsible for making parking a contentious mobility issue.

The expansion of the downtown area, along with increasing migration into neighborhoods near the central district and the transformation of the Loop into a 24-hour activity center, testify to Chicago's recent prosperity. This growth has been accompanied, however, by only a slight increase in total available parking spaces, while monthly parking permit prices have swelled in many cases to well over $200 per month. Many officials hesitate to call for policies to promote dramatic increases in parking availability out of fear that commuters would increasingly switch from public transit to driving, thereby worsening downtown traffic congestion and air quality. While the city has included parking space minimums for proposed redevelopment projects and approved construction of new garages, many residents and business owners feel that the apparent shortage of parking will only get worse in the years ahead.

Throughout Chicago's central business area, several private companies, such as Standard and Allright, are principal suppliers of parking space. These firms make substantial investments in facilities and compete fiercely with each other for the patronage of thousands of commuters. Nevertheless, the city, as both a major provider of parking and a policymaking body, continues to have a profound effect on the investment and pricing decisions of the private parking companies.

A commuter parking lot on Halsted Street provides convenient access to River North and the central business district.

The Department of Planning and Development described the city's parking dilemma in a 1998 study, which showed a marginal increase since 1992 in the total number of spaces in the Loop and the surrounding downtown area. This increase of 2,082 spots, to nearly 87,000, has failed to keep pace with growth in downtown office space and residential population. Having increasingly turned to the private sector to provide parking facilities since the 1970s, the city is relying on redevelopment projects to attract investment in new garages. In recent years, the city's increasingly unobtrusive stance has allowed private companies to become the primary suppliers of parking in response to growing market demand. At the same time, however, the city has refused to yield from its long-held position that on-street parking should be limited to approximately 900 meters and 150 free spaces throughout downtown—figures dating back to the temporary ban on street parking following the flood of 1992.

Resolving this issue has important implications for growing numbers of commuters and noncommuters. At present, half of the estimated 2 million people who visit downtown each day use public transportation. While both the Chicago Transit Authority (CTA) and Metra report rising trends in work-related trips, the number of nonwork-related transit trips is falling, suggesting that parking concerns will be especially significant to the future of downtown shopping and leisure activities.

The growth of restricted-parking zones has enormous implications for solving Chicago's parking problem.

The economic stakes involved have risen considerably over this decade as restricted- and permit-parking zones (e.g., resident only, valet, loading, night baseball) have spread across the city in response to rapid development. These zones, first introduced in the late 1970s to preserve space for residents in neighborhoods around downtown, have mushroomed to more than 500 citywide with 54,000 permits issued annually. These designated areas have become increasingly unable to accommodate the influx of affluent urban residents who desire convenient parking for personal automobiles.

At the same time, enforcing the restricted-parking zones has become an important revenue source for the city. Fine collection alone has risen from $20 million in 1989 to over $70 million in 1998. Accordingly, the city stands to lose significant income if neighborhood parking is reopened to nonresidents and other restricted-parking zones are reduced in scope.

Adding to the administrative complexity of parking regulation is the fact that several municipal departments maintain oversight of different aspects of parking policy. The Department of Revenue enforces fee collection, the Department of Planning and Development negotiates parking logistics with developers, the Zoning Board considers legislation, and aldermen promote development policy and respond to resident pressures. The entire debate easily becomes framed by self-interest and occasionally overlapping perspectives that can lead to disputes pitting neighborhoods against each other or residents against neighboring businesses.

Concerns for the future of downtown parking loom large on the urban planning horizon.

Even after commissioning a task force and conducting a detailed study of the prominent issues, Chicago will likely have the resources to provide only modest relief to the parking dilemma. The study, however, proposed more than 40 possible measures, including re-evaluating and setting a five-year life span for all restricted-parking zones, increasing parking sticker fees, opening school and church parking lots during evening hours, building more park-and-ride lots, and creating a trolley service in Lincoln Park.

The city evidently recognizes that mitigating its parking problems will require a willingness to experiment with new strategies. Among the more innovative ideas to be considered by the city in discouraging motorists from driving downtown are having merchants provide CTA passes to customers to promote transit use and promoting Transit Checks, a Regional Transportation Authority program that offers vouchers reimbursing businesses who pay commuting costs of employees. Another possible initiative is creating "parking innovation wards" where new ideas for easing the parking dilemma would be tested. These measures rise above the short-term responses that have angered many motorists, yet became increasingly common in recent years. The added capacity resulting from these measures could create as many as 10,000 new spaces. However, this will fall well short of predicted downtown growth and provide only a small step toward satisfying business interests and residents alike.

Although Chaddick's desire for Chicago to conduct comprehensive studies of the parking situation has been fulfilled, the challenge of implementing solutions that address the root causes of parking problems remains unmet. These studies suggest that the city must now proceed with a renewed sense of urgency. Officials must move beyond stopgap, interim measures and begin considering broad answers to prevent insufficient parking capacity from becoming a detriment to the municipal economy.

11

Adaptive Reuse of Vacant Structures

Adaptive Reuse of Vacant Structures and Underutilized Large Tracts

"Chicago is developing an increasing inventory of vacant schools, churches, and other institutional structures, as well as commercial loft buildings. Many of these large structures occupy prominent locations in the neighborhoods in which they are found. Can we prepare a rational policy for the appropriate reuse of these sites, and incorporate this policy into a new comprehensive plan?

"In addition, although the city is almost 100 percent 'built-up,' extensive tracts and strips of land remain either unused or underused, much of it owned by railroads and large corporations, especially utilities. This land should be considered by the city as potential sites for planned development, which could provide housing, appropriate mixed use, and usable open space.

"Corporate-owned vacant land could be acquired by the city through purchase, exchange, or even voluntary contribution to the project, which might result in a valid tax write-off for the property owner."

— Harry F. Chaddick

Contemporary Assessment

Chicago's streetscape reflects its longstanding role as one of America's foremost industrial cities—with relatively little undeveloped land awaiting commercial use. Therefore, as Chaddick suggests, the city has a strong financial incentive to adroitly reuse its land, structures, and natural resources in order to reap the benefits of economic expansion. In recent years, the city has explored many opportunities to reuse vacant properties, while remaining mindful of common planning goals for the larger community.

Evidence from three sites described below suggests that this remains an enormously complex and multifaceted task. These paragraphs summarize the experiences of the former USX South Works steel mill, the Montgomery Ward Chicago River property, and the Lakeside Press complex formerly owned by R.R. Donnelley—urban sites exemplifying the virtues of reclaiming vacant industrial parcels for commercial purposes.

The city's enormous South Works project illustrates the need to delicately balance the goals of an entire community with the more localized concerns of a particular neighborhood.

For nearly a decade, the South Works property was the largest dormant tract in Chicago available for redevelopment. Throughout this period, residents and officials from the surrounding community were dismayed with the lack of municipal action, and private sector leaders remained reluctant to invest in the massive and isolated site.

The first integrated rail mill ever constructed, South Works became an integral part of our region's economy soon after it opened in 1882. With a total area larger than the Loop—encompassing some 573 acres—it served for generations as a venerable symbol of our nation's industrial might. The enormous facility provided steel for some of Chicago's most important structures, from the Wrigley Building to the Sears Tower. U.S. Steel, the largest company in the world at the time, bought the facility in 1901 and guided its growth to a high of 20,000 employees in 1939. A victim of technological change that doomed many heavy industries across the country, USX Corp., the successor to U.S. Steel, closed South Works in 1992 and began preparing the site for sale. USX hoped to sell the lakefront parcel in one piece, but initially generated little interest in this scenario. Several commentators suggested the $85 million price was too high for a single developer to risk on a parcel 10 miles from downtown. Adding to skepticism was the parcel's questionable environmental history and poor roadway links, as well as its proximity to the lake and many active rail lines, which limited vehicular access.

As time passed, the city gradually became more proactive in the planning of the site—especially after USX was unable to find a prospective buyer for the land stretching from Rainbow Beach at 79th Street to 91st Street near the mouth of the Calumet River. Facing growing criticism from community groups and local politicians who felt the city was being too passive in encouraging the redevelopment of South Works, officials reiterated the long timeframe necessary for a thorough analysis of the site and all possibilities for a subsequent project.

The city's bold aspirations for the property became apparent when the Department of Planning and Development produced a 1998 report, *A Community Vision for Future Development,* that compiled the expressed wishes of area residents and business owners who had attended a series of workshops exploring the site's various possibilities. The city also hired a planner specifically to manage the redevelopment effort and began working with public and private sector organizations to assemble a plan for the site. In early 1999, the city released a pair of plans prepared by Skidmore, Owings and Merrill. At an estimated cost of $200 million for public improvements, both plans called for a mix of commercial and housing development, as well as the designation of 131 acres, or 20 percent, for public and park space.

One of the most notable innovations of these plans is their sensitivity to uncertainties in the market during the implementation stage. The plans provide two development scenarios that allow officials to respond to the changing market and alter the mix of residential and commercial land-use. However, the public investment projects, such as additional Metra stations, the rerouting of U.S. Highway 41, and other neighborhood improvements, are scheduled regardless of the eventual development scenario.

Optimism about the site's redevelopment rose in June 1999 when Solo Cup Co. announced the acquisition of a 107-acre parcel of South Works to build a $71 million manufacturing facility to be completed in 2001, bringing 750 jobs to the area. The agreement includes $16 million in assistance from the city ($14.9 million of this total from an approved tax-increment financing district) and $15 million from the state. When finished, area planners believe the site will be a model of thoughtful urban redevelopment.

The many incentives used by the city to promote this development package represent a shift in policy that will likely have significant repercussions for other parcels of underused land. The close cooperation between the city and state in fashioning a financial assistance plan, as well as the willingness of officials to work closely with outside groups, is worthy of considerable praise. Although the development of the entire 573 acres is expected to take 20 years to complete, the city's careful assessment of the site's advantages and disadvantages provides encouraging signs that the city recognizes the market opportunities for underused tracts within its jurisdiction.

The Lakeside Press complex poignantly exhibits the benefits of matching the needs of the private sector and the redevelopment goals of the community.

This project, along with the Montgomery Ward Chicago River property, indicates the city's willingness to take risks in fostering the reuse of old industrial structures that many considered functionally obsolete.

The Lakeside Press complex sits next to McCormick Place on Chicago's Near South Side.

The Lakeside Press complex is a Near South Side historical landmark that housed the printing operations of R.R. Donnelley & Sons Co. through 1994. Three years later, a developer purchased the three-building complex, including the printing plant that was one of the largest in the world and produced the venerable "Big Book" catalog for Sears, Roebuck, and Co.

Seeking to take advantage of proximity to the burgeoning South Loop and McCormick Place, plans initially called for the site to mix residential, office, and retail uses along with a hotel. Although questions remained concerning traffic patterns and the appropriateness of the project for the neighborhood, officials approved the 10-acre site at 22nd Street and Calumet Avenue as a tax-increment financing district in 1998. In May 1999, a private company, in turn, purchased the voluminous eight-story Calumet Plant building, an edifice widely considered one of the finest examples of industrial Gothic architectural style in America. The company announced its plan to oversee the conversion of the building into the nation's largest Internet equipment and telecommunications center for technology companies. The $250 million plan will culminate in the creation of the 1.2 million-square-foot Lakeside Technology Center. The original developer will pursue revised plans for a mixed-use project on the remaining four acres of the property, which also include the 80,000 square-foot American Book Co. Building.

Civic groups applaud the project for its potential to revive the area between McCormick Place and Chinatown, a community that has been largely overlooked during the current building boom. Officials herald the substantial community benefits associated with the project, including a $1 million donation from the developer for a park adjoining a new public elementary school that the city will construct nearby and a pledge to fill construction jobs with Chicago Housing Authority residents. The city also provided a $4.5 million subsidy for energy infrastructure improvements to the building, showing the desire to position Chicago as a prominent participant in the technology-centered modern economy.

The former Montgomery Ward & Co. site illustrates the potential for providing new roles for aging industrial structures in neighborhoods increasingly acclimated toward urban professionals.

A portion of Montgomery Ward's property along the North Branch of the Chicago River, including the landmark catalog building constructed in 1908, was recently purchased by a developer who is waiting for city approval to begin an ambitious redevelopment project. Included in the sale of the 23-acre site are 2.5 million square feet of buildings and more than 13 acres of vacant land planned for housing, office, retail, and restaurant uses.

The city considers the property's river frontage an important asset, one instrumental to its plans for improving public access to the open space adjacent to the waterway. The land's proximity to the Cabrini-Green public housing complex (whose buildings are also scheduled for redevelopment) has evidently not affected the negotiations. Indeed, the success of various renewal efforts east of Cabrini-Green is contributing to citizen confidence in the neighborhood's commercial vitality. Using the historic river area as a tool to attract private capital, the city is reaffirming its commitment to improving this troubled area.

These latter two projects symbolize Chicago's willingness to work tirelessly to promote anchor projects in neighborhoods that had fallen out of favor among developers. Seeing an opportunity to initiate projects that will spur further development, the city has successfully critiqued different development proposals before finalizing financial agreements with redevelopers. Officials have worked to minimize infrastructure expenses and take advantage of urban locations near suppliers, shippers, and a growing downtown pool of professional labor and services.

In summary, there is growing evidence that Chicago has prospered by concentrating on the most encouraging redevelopment opportunities. Nevertheless, as older neighborhoods depreciate in response to economic change, it is an opportune time for the city to seriously explore the virtues of moving toward an even more comprehensive strategy that promises redevelopment within the context of a larger plan. In this respect, a pervasive case can be made that redevelopment should be viewed as an integral part of a regional effort to strengthen the central city.

Chaddick's recommendation of methodically and imaginatively approaching redevelopment possibilities remains as relevant today as it was during the 1960s and 1970s, when heavy industry accelerated its retrenchment throughout the Great Lakes region. Only recently, however, has the city found ways to channel its limited resources into successful projects that raise hope for the metamorphosis of vast tracts of urban land.

12

Superblocks

Redevelopment of Older Urban Areas into "Superblocks"

"Sometimes it seems that Chicago is the prisoner of its grid system of streets and alleys. Back in 1830, prior to the incorporation of the city as a town, the first streets were laid out. Each was 66 feet wide. The north/south streets ran at right angles to the east/west streets. Alleys, 16 feet wide, slashed through the center of every block. Over the next 100 years, this grid pattern was extended throughout the city. It is not until you get to the suburbs that there is any major interruption in the pattern.

"The rectangular grid has been so predominant that it led one 19th century traveler to describe Chicago as 'the most right-angled city in the United States.' The city's near slavish adherence to a pattern of streets and alleys established before the inventions of the 19th and 20th centuries, such as automobiles, is just plain foolish.

"I believe we should get away from this grid pattern and that portions of the city should have what I call superblocks. These would be tracts of land any size larger than a block which would not be interrupted by streets or alleys. For example, one superblock might be one-half square mile. Along with single-family dwellings, this tract might include a school. The objective of creating such a block would be to enable city residents to live in a home environment that is largely carless, pedestrian-oriented, and very safe for children.

"With the absence of streets, much more space could be devoted to greenery. Recreational facilities could include a fieldhouse, swimming pools, basketball courts, golf courses, and softball diamonds. The lack of streets and alleys would free a substantial amount of land

which could be used for tennis courts, children's playgrounds, or dozens of other uses now prevented by the unrelenting grid pattern of streets and alleys. A system of cul-de-sacs could be installed to enable homeowners to park their cars in their own garages while, at the same time, not having thoroughfares crisscross through their residential area.

"Many suburban residents enjoy the amenities afforded by carefully thought-out uses of land. Why not make these amenities available to city residents? In low income areas, why can't a sizeable number of vacant lots and properties occupied by burned-out structures be assembled into nongrid tracts for development into land-uses providing housing, recreation, and aesthetic beauty? Must these types of land-uses be available only to the rich or upper-middle class?

"Up to now, we have lacked the imagination to try something like this. There is no end to the ways we can improve urban living by more imaginative use of land. Too many of us have pursued a rigidity of thinking that does not permit land-use creativity."

— Harry F. Chaddick

Contemporary Assessment

The role of neighborhoods in the lives of urban residents evolves gradually in response to new technologies and cultural norms. From the confined urban centers that emerged during the Industrial Revolution to the decentralized suburbs emerging after World War II, the shape and physical qualities of neighborhoods have contributed to the nation's commercial vitality and its distinctive social fabric. Both sociologists and urbanologists support Chaddick's point that the configuration of a neighborhood has profound implications for the extent and quality of interaction among its residents.

Many share Chaddick's larger vision of boldly changing the streetscape to improve the quality of life in various parts of the city. Several commentators have championed similar plans to convert streets into cul-de-sacs, establish new parks, and increase pedestrian activity.

The following paragraphs briefly review several such projects.

Although the concept of superblocks has not been directly applied in Chicago, many ancillary tools have helped strengthen neighborhood cohesiveness.

Many residents of Chicago are unaware of the bold policy initiatives that are underway to bolster neighborhood interaction. Among the many significant initiatives that support increased community interaction are Chicago Transit Authority (CTA) superstations, empowerment zones, the Strategic Neighborhood Action Program, and the Chicago Alternative Policing Strategy. Also, the increasing national popularity of traditional neighborhood design and growing awareness of "walkability" initiatives serve to extend Chaddick's idea.

The discussion surrounding the creation of CTA superstations is the result of an urban renaissance that is drawing many new residents to Chicago's historic neighborhoods. It also reflects a growing municipal commitment to integrating rail transit services into the economic development process. Using superstations as a catalyst for neighborhood improvement, the CTA is consolidating its disparate bus services at these sites and making them focal points for transit-oriented developments. Chaddick would likely applaud renovation efforts, such as the rehabilitation of the Green Line elevated rail system, which is providing the impetus for significant new commercial and residential activity.

Three Chicago neighborhoods, in areas generally referred to as the West, Lower West, and Near South Sides, have been designated empowerment zones under the federal Empowerment

Zone/Empowerment Community program. This highly regarded initiative gives stakeholders in troubled neighborhoods access to substantial financial resources to work together to craft strategies to revitalize their communities and create job opportunities. By 1999, local efforts leveraged funding from block grants, tax incentives, and local public and private sources to start nearly 100 projects in neighborhoods designated as empowerment zones. Although the program's management and the sometimes-disappointing level of resident participation in the decision-making process have drawn criticism, the program has been applauded by many who feel it is a promising first step in reversing the fortunes of some of the city's poorest neighborhoods.

The Strategic Neighborhood Action Program is another notable effort to concentrate a wide array of local resources within a neighborhood. Through the leadership of several city departments and private investors, the program aspires to improve infrastructure, demolish abandoned buildings, and assist in commercial development. This program has initiated projects in seven neighborhoods, ranging from Albany Park to Southeast Chicago.

Finally, the Chicago Alternative Policing Strategy is broadly supportive of the movement to promote neighborhood autonomy. This program began in selected neighborhoods around Chicago and has recently been expanded citywide. As active participants in creating a more secure neighborhood, residents share a renewed sense of control over their community. The Chicago Alternative Policing Strategy initiative encourages citizen involvement to foster neighborhood cohesion and helps develop a common vision for the area. It serves to galvanize residents and encourages them to assume responsibility for their community's social and physical character.

A vacant West Side parcel that is centrally located and transit-accessible.

Rising appreciation of the psychological and sociological effects of neighborhood design has led to strong grass-roots support for policies instilling a stronger sense of community throughout the city and region as a whole.

In recent years, many development projects have incorporated design characteristics commonly associated with the New Urbanism movement. Intended to reduce the social isolation of typically decentralized suburban neighborhoods, this traditional development style incorporates more interactive features—including more open space, short setbacks from the sidewalk, detached garages, and greater visual variety than is normally found in single-family housing developments.

These ideas share a consistent set of values with the movement to promote walkable neighborhoods. Several prominent organizations have joined the Partnership for a Walkable America, which has led national campaigns to make urban areas safer and more accessible to residents and pedestrians. Recognizing that achieving these goals will require more emphasis on transit-oriented development and safety considerations, the U.S. Department of Transportation has added pedestrian access to its capital project planning process. These efforts are supported by evidence showing that a neighborhood's walkability and its proximity to retail outlets are often prominent factors in the decisions made by families to relocate to the suburbs rather than remain in the city.

These are just a few of the contemporary approaches that illustrate the need for neighborhood-specific investment and planning to improve the character of community life. Although the benefits of these initiatives are evident, the city faces many obstacles in pursuing them on a truly citywide scale. Achieving Chaddick's larger vision might remain a distant possibility, but the spirit of his ideas is apparent throughout Chicago.

13

Effective Use of Park Land

Policies and Action Leading to More Effective Use of Chicago's Park Land

"The city's parks, most of them designed and laid out in the late 19th century, are, for the most part, essential contributors to the aesthetic and recreational enjoyment of the city's residents. Many of these parks are the work of the great landscape architect Frederick Law Olmsted and his partner, Calvert Vaux. In some parts of the city, their plan resulted in a greenbelt with broad, tree-lined boulevards. This greenbelt on the South Side consisted of Jackson and Washington Parks linked by the Midway Plaisance. On the West Side are Douglas, Garfield, Columbus, and Humboldt Parks. The sizes of these six parks range from 135 acres in Columbus Park to 543 in Jackson Park. Each has played a vital role in the life of Chicago. But problems remain.

"In the approximately 100 years since the creation of these parks, there have been many changes. In some instances, and for many reasons, there has been a deterioration of the surrounding neighborhoods. What is more, originally there may have been a mistake in making some of these parks too large. Excessive size can lower usability, and a decline in usability can result in near-deserted acreage. This is what has happened in some instances. One of the results has been that some portions of these parks are definitely unsafe for public use.

"This concentration of land in parks larger than 50 acres in size has been accompanied by the creation of too few small and properly located neighborhood parks, five to 10 acres in size.

"I suggest we consider an innovative land exchange program in which the Park District would set aside 20 or more acres of underused land from each of Chicago's major parks and utilize the land by constructing on it low-cost single-family and two-family residences. The adjacent neighborhoods would benefit from the elimination of underused and sometimes forbidding park land. The owners of the new low-cost housing also would benefit from having their homes located in park-like settings which no longer would be deserted, but would be filled with the vitality of the new residents. Appropriate security measures could be provided for these new housing clusters.

"A second phase of my proposal is that the 20 or more acres utilized for housing in each of these large parks would be accompanied by the creation of an equivalent amount of new park acreage in the same general areas of the city. These new 'mini-parks,' 10 acres or smaller in size, would be built on vacant lots and properties now occupied by derelict structures. These parks could be built in any configuration that would utilize the vacant lots and be convenient for nearby residents. If each new park were to be as large as 10 acres, this would mean the addition of at least a dozen parks to compensate for the 120 or more acres in the large parks where low- to moderate-cost housing would be constructed.

"This 'new housing-new parks' proposal can be accomplished through the coordination of the city government, community organizations, financial institutions, labor unions, and developers, and should become an integral part of a new comprehensive plan for Chicago."

— Harry F. Chaddick

Contemporary Assessment

The rich tradition and spacious dimensions of many of Chicago's parks are simultaneously advantageous and disadvantageous to the city's residents. Although the bold aspirations of early park designers left our community with a considerable amount of open space, the size and location of this

The gardens of Grant Park on South Michigan Avenue augment downtown lakefront attractions. *(Credit: Willy Schmidt/City of Chicago)*

space create enormous challenges for those charged with maintaining the park system and meeting evolving neighborhood needs. Commentators such as Chaddick understood the urgency of a fundamental restructuring of Chicago's park space. He suggested constructing affordable housing on underused sections of large parks while creating an equivalent amount of park acreage on vacant or underused lots in the same neighborhood.

The paragraphs below summarize several important issues that underlie this ongoing civic discussion.

Recent administrative changes at the Chicago Park District raise the possibility of dramatic improvements to park land and recreation acreage throughout the city.

As a legal entity separate from the municipality, the Park District maintains its own budget and taxing authority. While the district has a long and distinguished history, it has faced allegations of political patronage over the years. Some have accused the Park District of being unable to respond to the changing times and evolving needs of Chicago's residents. After maintenance and budget problems in the early 1990s profoundly affected the Park District, the city initiated strong reform measures. By 1994 it had canceled hundreds of capital projects and eliminated a quarter of its workforce, marking the beginning of a new era for the Park District.

Many innovative Park District efforts emerged from this difficult transitional period. One notable outcome is the CitySpace program, which began in 1993. Through the leadership of the Chicago Community Trust, this two-year project brought together personnel from the Park District, Cook County Forest Preserve District, the city, and other institutions to draft a comprehensive plan to acquire and develop new park land. Published in 1995, the plan is sensitive to the finding that 61 percent of Chicago residents live in neighborhoods where parks are either overcrowded or not easily accessible. The plan recommends that each of Chicago's neighborhoods include a minimum of two acres of park land per 1,000 residents—a goal the plan's organizers hoped to achieve in 10 years. Nevertheless, only 38 of the city's 77 neighborhoods presently meet this suggested minimum.

Among the 130 proposals included in the CitySpace plan are measures to convert concrete and asphalt lots that surround many schools to grass-covered recreation areas. Since the plan's release, many schools have used a subsequent school board allocation of $600 million in capital improvement funding to help redesign schools to reflect the "campus-style" facilities commonly found in the suburbs. The plan also recommends the city revise the Tax Reactivation Program, making it easier for community groups to become the stewards of tax-delinquent and city-owned vacant lots, which have generally been earmarked in the past for affordable housing. Another significant part of the plan is the creation of new trails, greenways, and wilderness preserves along abandoned railroad corridors and inland waterways such as the Chicago River and Lake Calumet wetlands.

Significantly, the CitySpace plan also focuses attention on several smaller-scale projects such as providing landscaping around government buildings, adding more green space to streets and medians, and supplementing Park District efforts with community involvement projects to help develop vacant lots. Intended to serve primarily as a blueprint for future open space efforts, this program awards funding on a case-by-case basis and includes several long-term projects conducted in cooperation with other agencies.

Citywide, Chicago has about 4.1 acres of open space per 1,000 residents and ranks 18th in a 1997 study exploring the amount of park land in major cities. Communities with ratios greater than Chicago include New York City, San Francisco, Cleveland, Boston, Dallas, and Los Angeles.

Following the release of the CitySpace plan, the Park District joined the city and Forest Preserve District in an unprecedented initiative, forming a nonprofit organization known as NeighborSpace. This entity strives to acquire and insure smaller park sites whose diminutive scale is inappropriate for the Park District or the Forest Preserve District.

The municipal government is adroitly using a variety of financial tools to induce developers to participate in the creation of open space.

Seeking to capitalize on the robust Chicago housing market, officials crafted a 1997 ordinance imposing "impact fees" to help pay for new park land. These are similar to the fees imposed in many suburbs requiring developers to help offset the cost of providing infrastructure for new residents.

The Chicago ordinance provides a mechanism to encourage developers of new housing units and rehab projects to provide open space as part of their developments. If unable to do so, they contribute to a fund to develop parks in the same neighborhood. The fees are based on the square footage of the living area—from a nominal fee for affordable housing to over $1,200 for homes with greater than 3,000 square feet. Though concerns remain about the possibility of escalating or additional development fees in the future, most analysts support this compensatory measure.

The rejuvenated and decentralized Park District has made notable strides over recent years in increasing its capacity to efficiently serve the residents of Chicago. The Park District is working in conjunction with other agencies to initiate new projects and rectify remaining managerial problems. Although Chicago will likely continue to have a low national ranking in the amount of park land per capita well into the new millennium, nearly 300 acres of park space have been added in the past six years. Chaddick, along with many other prominent civic figures who witnessed the growth and evolution of Chicago and its recreational needs, would look with approval toward these recent initiatives and implore leaders to reject the natural tendency to merely perpetuate the status quo which does little to shape our parks for the growing complexity of urban life.

14

Efficient Zoning and Construction Approval Process

Streamlined and Expedited Process of Zoning and Construction Approval

"Criticism has been leveled at the Chicago building and zoning permit process as being far too lengthy, confusing, and unwieldy. Can the city introduce a one-stop or fast-track permit process to eliminate delays? Is there merit in allowing for (or even requiring) a pre-application conference with a 'development coordinator' who would review the developer's intent, explain the codes and procedures, and be assigned to lead the applicant through the process?"

— Harry F. Chaddick

Contemporary Assessment

Largely unheralded in Chicago's urban development history are the profound—and sometimes deleterious—consequences of its municipal zoning and building code approval system. Some contemporary commentators consider the system one of the country's most antiquated and characterize it as inefficient. They purport that it often frustrates developers, requiring that they deal with delays and seemingly arbitrary rulings.

Chaddick's sentiment about the city's construction approval process is today shared by a new generation of real estate professionals. Recent events, however, promise to significantly affect both the Department of Buildings and the 1,200-page Chicago Building Code.

Several important aspects of this continuing discussion are briefly described below.

The growth of the construction industry in Chicago places the existing regulatory and bureaucratic system under severe stress.

The combined effects of economic expansion, low interest rates, and rising demand for housing and business locations around the city have led to a boom in residential and commercial construction. Exemplifying these prosperous times, the number of permits issued by the city rose from 21,000 in 1996 to 25,000 in 1997 and 30,000 in 1998. One unfortunate consequence of such growth is the vocal protests from developers and homeowners about the lengthy waiting periods for receiving construction approval.

Chicago was the first U.S. city to adopt a building code in 1875, four years after the devastating Great Fire. Although the city overhauled the code on several occasions, it has made only minor changes since 1957, when the last major rewrite took place. Calls to update a code cluttered with superfluous language that addresses such obsolete issues as the safety of dumbwaiters (small, manually operated elevators used primarily in bygone eras to send items such as food or trash between floors) gradually escalated as the city modernized. By the late 1990s, complaints reached a feverish pitch as developers reported delays of up to four months for a building permit. Chicago's cumbersome code and the unwieldy building department approval process, which often consumed far more time than the two to four weeks typical in other cities, reportedly had detrimental effects on construction projects around the city. In recent years, construction firms have responded by turning to permit expediting services—firms charging close to $80 per hour solely to handle the permit issuance process for developers.

The Chicago Department of Buildings, with a staff of more than 400 employees and a budget of $27 million, remains the guardian of the city's building code policies. A 1994 study outlined instances of wasted resources and needless duplication within the department. Unfortunately, many modernization efforts intended to improve departmental efficiency fell short of expectations. One such effort was a computer upgrade to digitally scan blueprints, thus allowing inspectors from different departments to review them simultaneously. When the blueprint images were reduced to fit the department's 17-inch monitors, the images became so small that officials occasionally sent plans back to potential developers to fix "errors" that were actually nonexistent.

Preliminary evidence suggests that recent improvements are helping to alleviate approval delays.

After enduring nearly relentless criticism for its ponderous permit approval and inspection process, the city marshalled the political will to overhaul the embattled building department. Concerned about the potentially adverse effects of delays on local construction, the current mayoral administration appointed a new departmental commissioner with a reform mandate. New management staff from the public and private sectors arrived at the end of 1998 with a coordinated program to simplify the permit process and decrease bureaucratic inefficiency.

Reports from officials and industry stakeholders provide room for optimism about the benefits of these administrative changes, which have reduced the time necessary to receive permit approval. Although this period can vary greatly depending on the magnitude of the project, commentators note generally shorter delays since the restructuring took place.

Rehabilitated buildings in Lincoln Park have benefited from Chicago's updated approvals process. *(Peter J. Schulz/City of Chicago)*

Still, many analysts feel that a further restructuring of the building department is overdue. Toward this end, the city is taking definitive steps to address operational issues within the department and coordinate the permit approval process. Hoping to create a one-stop permit process, officials have relocated representatives from all departments having approval duties to contiguous floors of City Hall and opened satellite offices around the city to more conveniently serve homeowners. The department also has printed a booklet describing all the permit approval requirements and has bypassed the troublesome computer software. New software now applies bar codes to all plans to track their progress and staff performance. In addition, the department realized that it returned nearly half of all new applications due to missing information; in response, all new applications are now given a preliminary check for completeness on arrival before moving to the detailed inspection phase.

The department is also allowing additional self-certification for projects built from standardized blueprints, thus eliminating uncertainty about the quality of design. Taking a cue from the private sector, the department has begun to provide its own "expedited handling" option for developers willing to pay an additional fee by contracting with outside companies to review selected projects. These changes, along with others implemented since 1998, have served to improve overall performance of the department.

For the first time in more than 40 years, fundamental changes to the content of Chicago's building code are a distinct possibility.

In 1999, the mayor surprised many observers by announcing that the department's new management will orchestrate a rewriting of the entire code. Although entrenched interests have traditionally stalled attempts to modify the regulations, some insiders believe that a rewritten code could be ratified soon. Diverse committees of field professionals are presently working to adapt widely utilized model national building codes to Chicago.

Minor revisions over recent decades have done little to reduce what many commentators claim to be inflated construction costs attributable to the dated code. Housing groups, for instance, cite evidence showing the cost of a single-family home in Chicago is 20 percent higher than in the suburbs, a cost these organizations attribute largely to regulatory issues. The new code could offer potential savings to developers and provide further incentive for the building department to streamline its operations.

All of these developments suggest that the provincial battles that have traditionally characterized the construction industry's relationship with the city appear to be slowly ending. Fortunately, many parts of the procedural bottleneck that analysts considered a threat to the construction boom have been addressed through reorganization and policy changes.

At the same time, many suburban codes remain insensitive to the growing need for affordable housing among middle- and lower-income constituencies. Advocates of such housing are persuasive in their claim that most suburban codes unnecessarily raise construction costs by discouraging or preventing the use of many building materials and technologies.

They find their views supported by *Metropolis 2020,* the regional planning document completed in 1998. The report calls for a statewide building code that would further simplify and centralize construction and zoning approval decisions—a recommendation whose implementation appears to be years away.

Chaddick, who spent many years trying to standardize development controls, would likely applaud such a lofty endeavor.

15

Comprehensive Plan Implementation Methodology

A Comprehensive and Coordinated System of Plan Implementation

"Any serious planning effort must be accompanied (or closely followed) by a companion piece—a well-thought-out and coordinated implementation system. When we think of plan implementation, we usually identify it with 'zoning,' and perhaps the zoning ordinance is the primary device for ensuring proper effectuation of the plan. But, zoning is not the only means of achieving the goals of the plan.

"A coordinated implementation system for Chicago should include the following:

- *A capital improvements program and budget. This should reflect projected expenditures for public facilities and infrastructure.*
- *Subdivision regulations. Although the city is almost completely platted and built upon, there will be continuing major redevelopment projects that require effective resubdivision and innovative lot layout design.*
- *A zoning ordinance. Chicago's ordinance has not been comprehensively revised since 1957. It is obsolete and out of step with modern land-use needs. We need a comprehensive re-examination of the present zoning ordinance, leading to a modern code working in concert with the new comprehensive plan. Special attention should be focused on the planned unit development concept, the regulation of nonconforming structures and uses, and sign and visual display controls.*

- *Building and rehab codes. Chicago's building code has been described as archaic and contrary to current construction techniques and the use of modern materials.*
- *Continuing maintenance codes. These regulations are designed to ensure continuing maintenance of development at safe code levels.*
- *Economic incentive programs and provisions. These include enterprise zones and tax-increment financing districts.*
- *All pertinent county, regional, state, and federal programs, and funding sources.*

"It is essential that these various programs, regulations, and codes be synchronized to work in concert to put the city's comprehensive plan into effect. Just as critical to success is the periodic review of the implementation system and its many components. Perhaps there should also be a mandatory regular joint evaluation of the plan and the implementation system.

"Devising an optimum plan implementation system for Chicago may be difficult but certainly is worthy of the effort."

— Harry F. Chaddick

Contemporary Assessment

Like most major cities, Chicago has extensively studied various urban problems only to discover that, in many circumstances, it lacks the tools and resources necessary to completely eradicate them. The city and other public and private agencies have assembled a veritable mountain of literature espousing new and innovative planning measures for improving urban life. Unfortunately, several of these plans, quite literally, have been relegated to the shelves of municipal and agency libraries—with little or no apparent impact on the community.

Chaddick believed strongly in linking planning efforts with implementation measures that would translate into effective and thoughtful municipal development. He would find it encouraging that the many area-specific plans that have emerged in recent years are sensitive to the difficulty of implementation in the city's volatile political environment.

The paragraphs below offer a summary of several pertinent considerations in this larger discussion.

The gap between analysis of municipal problems and corrective actions is attributable in part to the fragmented nature of governmental decision-making.

As long-time residents of Chicago are well aware, the city's politics tend to be inordinately complex and characterized by tremendous inertia, which renders meaningful change difficult. The City Council is often preoccupied with the provincial concerns of its diverse membership. The state government has historically assumed a relatively passive role in the promotion of evolving urban planning initiatives. Moreover, governance throughout the region is growing increasingly fragmented as the metropolitan area expands, with more than 1,200 units of government and numerous councils of government playing a role in the formation of policy.

One telling illustration of the difficulty facing the city's legislative process: the City Council has repeatedly failed to act upon calls for a comprehensive revision to the city's zoning ordinance, which has not been significantly altered since 1957. Chaddick would likely be disheartened by the lack of progress in this area, which requires citizens and businesses to conform to a code that is now widely considered to be outdated. It is encouraging to note, however, that the council began considering in 1999 serious proposals for updating the ordinance. This positive step, as well as the strength of the economy and other initiatives, suggests that a window of opportunity may be opening in which

The John Hancock Building is a focal point for comprehensive planning along celebrated North Michigan Avenue. *(Credit: Peter J. Schulz/City of Chicago)*

comprehensive planning and land-use controls in Chicago may receive the attention they so richly deserve.

The current mayor's efforts to gradually strengthen the city's management team is leading to renewed optimism about the city's planning process.

Mayor Richard M. Daley has been diligent in his efforts to staff the city's various departments with highly educated and experienced professionals who understand the enormity of change affecting the municipal landscape. This management team recognizes that incremental steps will not be enough to abet the city's transition from an aging industrial city into a leading finance and service center. This is exemplified by the city's ongoing efforts to privatize many tasks long provided by city employees, thus lowering the costs of providing service.

In a similar vein, *Metropolis 2020,* the Commercial Club of Chicago's 1998 comprehensive planning initiative, offers a framework for implementing new planning strategies for the entire metropolitan Chicago region. This publication advocates the creation of a 15-member executive council that would oversee this process over 20 years. The initiative calls for the establishment of a regional organization that would consolidate the responsibilities of several of the area's principal planning agencies. At the local level, area municipalities would share some of their authority over land-use with this organization, which would integrate many previously uncoordinated municipal functions within the context of a participatory regional dialogue. For these reasons, of course, some stakeholders are unlikely to embrace the plan in its entirety. Nevertheless, *Metropolis 2020* merits acknowledgment for its considerable attention to implementation.

These and other landmark initiatives address squarely many of Chaddick's concerns. They suggest that enlightened civic dialogue could allow our region to come to terms with problems adversely affecting businesses, residents, and nonprofit institutions for many years to come.

Public Awareness and Professional Education Issues

16

Continuing Education for Development Professionals

Formal Programs of Continuing Education and Professionalism in the Development Fields

"The professionals responsible for designing our cities and structures and ensuring proper development and continuing maintenance are to be saluted for their many achievements. However, I contend that the results of their efforts should be on an even higher level, and would be if these urban planners, architects, engineers, inspectors, and other land-use professionals were required to attend periodic courses and seminars to continue their education and keep them informed of the latest techniques and developments in the field.

"Too many land-use practitioners seem to end their education with graduation from college. They accept a position and isolate themselves from the constantly evolving development scene.

"It is essential that our shapers and guardians of the environment continue to grow and experiment and interact with their colleagues from other firms and related disciplines. The professional organizations and our local colleges and universities should assume the responsibility for establishing continuing education programs for land-use professionals in the Chicago region."

— Harry F. Chaddick

Contemporary Assessment

The avenues of formal education among planners have remained largely unchanged since Chaddick expressed these views. Reputable sources of training, such as universities,

Grant Park's Buckingham Fountain stands as a forward-looking expression of the virtues of lakefront and downtown planning dating back to 1927. *(Credit: Willy Schmidt/City of Chicago)*

government agencies, and nonprofit organizations, are perpetuating their traditions of offering a diverse array of programs. Nevertheless, our region has been only partially successful in bridging the chasm between formalized education and the practical application of emerging planning concepts.

The points below briefly summarize several topics relevant to this discussion.

Two Illinois universities with formal planning programs constitute the primary source for planners to develop formal skills.

Two schools offer accredited planning degrees: the University of Illinois at Urbana-Champaign (with undergraduate and graduate programs) and the University of Illinois at Chicago (with a graduate program). Both have exemplary reputations for scholarly research on the dynamic theoretical aspects of the urban economic and social arena.

In addition, several other institutions in the Chicago area offer programs such as public administration, urban studies, and geography, including DePaul University, Governor's State University, Illinois Institute of Technology, Northeastern Illinois University, Northern Illinois University, and Roosevelt University. These schools allow students to gain the professional acumen necessary for agencies working extensively with development and planning-related issues.

Several consultants and municipal officials have expressed concerns about the success of prominent planning schools throughout the region and country as a whole in imparting the specific skills necessary to meet the needs of today's planners, who require a diverse and technical base of knowledge. The traditional academic perspective tends to focus predominantly on conceptual and social dimensions of planning, rather than on fostering practical tools required to contend with the complex dilemmas facing the contemporary municipality or other units of government.

Some degree of abstraction is a necessary component in any scholarly environment. Across the country, however, a separation appears to be growing between the theoretical concerns of academicians and the common problems facing working professionals. Such questions require interdisciplinary reasoning that often defies conventional scholarly approaches. Some analysts believe more rigorous training in the physical dimensions of planning, such as site design, land management, and the legal ramifications of alternative approaches to zoning, would help remedy this apparent didactic shortfall.

A promising initiative to bridge this educational gap includes periodic training sessions intended for the practicing professional.

Organizations playing a role in this process include various chapters of the American Planning Association, the Northeastern Illinois Planning Commission, the Metropolitan Planning Council, and other respected institutions. These groups work to build coalitions of planners and stakeholders, and play an essential role in the broader pursuit of thoughtful urban development. Professional education programs, however, constitute a single component in their large and complex agendas.

In an attempt to fill the need to provide training on several of the more technical aspects of urban planning, especially the nuances of development control (such as sign control, building codes, and zoning), the Chaddick Institute for Metropolitan Development began in 1995 a technical workshop series. These workshops consist of half-day sessions led by experienced professionals from the field. More than 100 municipalities in the region have sent representatives to these events.

Curricular changes at universities could help impart greater knowledge of contemporary planning problems.

For this process to reach its fullest potential, academic institutions and professionals should consider curricular modifications that anticipate practical needs and should encourage coursework after graduation that speaks directly to modern planning dilemmas. This training could be accomplished through coursework tailored to regional needs or informal not-for-credit programs that require a sustained time commitment from participants. Planners and their employers could take a lead role in this process by dedicating time and resources toward the development of these programs. They could prioritize key areas requiring supplemental educational attention.

Chaddick stood behind his convictions about the need for more education by creating, together with his wife, Elaine, the Chaddick Institute, which works in cooperation with DePaul's Public Services Graduate Program to meld scholarly and pragmatic endeavors into a cohesive curriculum that meets the needs of government and nonprofit officials.

17

Continuing Education for Officials and Municipal Staff

Periodic Training, Retraining, and Refresher Courses in Basic Planning, Zoning, and Development Areas for State, County, and Municipal Officials and Staff

"I remain optimistic about the future of our cities despite the many problems and obstacles facing us. It is optimism derived from the belief that America has the brightest and most eager-to-learn young people in the world. Cities and villages (along with counties and the state) need young men and women who have been schooled in the appropriate use and maintenance of urban land. Such schooling is needed by governmental employees all the way to and including mayors.

"How can any municipality continue to entrust its planning, zoning, and building decisions to legislators and administrators who know little about these subjects and have little or no information available to them? How can cities continue to allow the construction of billions of dollars in real estate developments to be determined by staff who have never taken courses or workshops in zoning laws, construction procedures, or the composition of building materials? Knowledge about building and the use of our precious urban land is as important to the life of our cities as the knowledge of law and medicine.

"Everything relating to urban land, from its grading and drainage to the design and construction of the buildings which occupy it, now in the hands of largely untrained personnel, must become the responsibility of well-trained technicians.

"Resolution of the problem described above should be a simple matter. We must offer convenient, low-cost, and relatively quick training

programs and basic education sessions to government officials and staff; perhaps even very basic half-day workshops conducted in the municipal or county chambers.

"The ideal mechanism would probably be the formation of a joint effort between the Illinois League of Municipalities, the pertinent professional societies, and local colleges and universities. This type and level of cooperation can work wonders!"

— Harry F. Chaddick

Contemporary Assessment

Unlike decades past when land-use planning required a relatively narrow band of technical knowledge, today's development process requires participants to maintain a wide range of professional and entrepreneurial skills. Stakeholders in this process now include municipal governments, economic development organizations, social welfare agencies, neighborhood

Chicago's Navy Pier is an emblem of the potential contributions of intergovernmental planning to reclaim underused civic assets. *(Credit: Peter J. Schulz/City of Chicago)*

groups, and state and federal officials—all approaching spatial development questions from different vantage points. Nevertheless, to assure congruent development policy, all stakeholders should have a basic understanding of urban planning concepts.

The following paragraphs point out several key components of a larger dialogue on this topic.

Among those in greatest need of an introduction to the elementary tools of urban planning are citizen committees serving municipal units of government.

These groups include appearance review committees, which oversee the application of municipal codes related to aesthetic qualities of a community; local boards of zoning appeals, which evaluate zoning requests; city plan commissions, which review and advise on planning proposals; city councils, which create policy affecting the land-use and development processes; and economic development commissions, which establish priorities for job creation as well as commercial and industrial expansion. These and other participants must collaborate with planners in many areas to achieve their mission.

Similar issues confront private sector and nonprofit stakeholders such as developers, chambers of commerce, and community-based organizations, which also struggle with elementary tenets of planning that profoundly affect the evolving patterns of land-use. They must frequently devote considerable time to simply understand the language of planning and development control. The absence of easily understood materials for public inspection often leaves large segments of the population unable to participate in the larger planning decision-making process.

Over the years, many attempts at filling this informational gap have met with partial success. Most notably, the University of Illinois at Urbana-Champaign sponsored the Zoning Institute, which provided a forum for municipal officials and other stakeholders to acquire an orientation to the principles of land-use oversight. This meeting was typically held over a weekend, allowing sufficient time for detailed discussions about the resolution of common problems. Increasingly prohibitive expense led to the termination of this program by the early 1990s. Although Governor's State University continues to offer courses relating to public administration and Northern Illinois University recently opened a new campus in Naperville with a public administration curriculum, the need for more instruction, especially in urban planning, has remained urgent in many parts of the region.

Several current initiatives are helping to fill the gap.

The Illinois chapter of the American Planning Association has reinstated, on a smaller scale, training sessions at periodic conferences held throughout the state. The Northeastern Illinois Planning Commission, as a federally mandated metropolitan planning organization for the region, also remains committed to providing educational opportunities to its constituents. These efforts, in addition to those offered by other agencies, could provide the foundation for a much larger-scale series of programs to disseminate information about prominent planning issues that may have eluded many public servants.

Other attempts to raise public awareness of planning concepts also merit acknowledgment. The Campaign for Sensible Growth, under the direction of the Metropolitan Planning Council, seeks to educate both private and public audiences about the implications of different approaches to development control. DePaul University now offers a Certificate in Metropolitan Planning that allows professionals from all sectors to expand their knowledge. Various other subject-specific programs also enhance participants' planning skills in many diverse areas. The Northeastern Illinois Planning Commission's training calendar serves as a widely circulated compilation of opportunities available throughout the region.

Nevertheless, proponents of increased training have been largely unsuccessful in persuading many stakeholders to seek instructional assistance. Regrettably, it may take years to convince those in greatest need that they should dedicate time and resources to their continuing education.

Incremental methods of addressing this issue could range from municipalities requiring formal training for their staff members to professional and labor associations making continuing education a condition of membership. Given the time and resource constraints of the modern municipal and agency staff, a multifaceted program that offers a variety of formal and informal educational options is perhaps the most logical approach. The academic community could play an important role in this process by providing a neutral forum for the discussion of contentious issues, as well as sponsoring educational programs designed to attract a wide audience. An academic-professional partnership with a long-term commitment to the needs of the field could serve as a foundation of more enlightened urban and suburban planning.

Unfortunately, as land-use professionals face increasing demands on their time, they must be willing to look beyond immediate concerns if such partnerships are to come to fruition. Chaddick's ultimate goal of fostering wide participation in training by citizens and staff has yet to be realized, though many officials may echo his wishes. In the interim, the organizations currently offering training opportunities deserve recognition for their efforts to reach public servants whose professional responsibilities grow more complex with each passing year.

18

Cooperation Among Land-Use Professionals

Promotion of Greater Cooperation Between Land-Development Professionals and Practitioners

"My seven decades of involvement in planning, zoning, real estate, development, and construction have brought me successes, challenges, and some disappointments. My various careers have allowed me to bridge the gap between land-use professional and developer-entrepreneur. I've worn both hats—planner and zoning specialist at times and real estate agent-developer-entrepreneur at other times. It is this unique dichotomy of vocations that allows me to focus upon one of my biggest disappointments.

"That disappointment would be the all too common (and wasteful) attitude of 'us against them,' or planner-zoner-architect-engineer (especially those in municipal employ) versus real estate agent-developer-entrepreneur. You would think a common cause would generate cooperation. The goal of everyone involved should be the best possible design and development for the community.

"We must find a way to generate a greater spirit of cooperation between those who propose and provide new development and those who review and approve such proposals. Perhaps we can begin with an annual joint conference that would bring together the various factions and provide a forum for old gripes and new ideas. Another possibility would be a joint council that could be empowered to offer workshops, seminars, and cross-training sessions, and perhaps even mediate disputes."

— Harry F. Chaddick

Contemporary Assessment

The evolution of land-use control over the last several decades is forcing developers, planners, architects, engineers, real estate agents, appraisers, and units of government to come to terms with their frequently overlapping responsibilities. Unlike years past when conventional zoning practices dictated patterns of development, today's land-use management is often characterized by planned unit developments, creative partnerships, and risk-sharing techniques such as tax-increment financing districts. This shifting paradigm, which Chaddick would have encouraged, requires more frequent and extensive communication and the need for innovative approaches to increasingly complex problems that affect many in the field.

The three points below offer perspective on these significant professional concerns.

At present few venues exist for various stakeholders to interact with professionals who all share the common goal of promoting efficient land-use.

Too often, provincial issues evolve into disputes that limit the possibilities for growth and frequently become simmering "turf wars." This contributes to the prevailing lack of public confidence in the development community and perceptions that the land-use development process is inordinately political and confrontational.

Prevailing stereotypes about developers and government agencies often do little to encourage a productive exchange between these groups. Developers frequently find themselves cast in an unfairly negative light despite efforts to address civic and social concerns. Government agencies are regularly portrayed as unimaginative and unresponsive bureaucracies. Such generalizations serve only to limit possibilities for cooperation and discourage the spirit of enlightened land-use decision-making. New and traditional gatherings of field professionals offer more than a chance to share and receive information, they provide the formal and informal interaction that is the basis of any future cooperative programs.

An important starting point in the process is for municipalities to reach out to the development community for assistance in modernizing municipal regulation.

Frequently, policy changes are made without considering those profoundly affected by these decisions. Examples include modifications to building codes, commercial zoning, sign control, and industrial performance standards (such as those governing noise and air pollution and other external effects of industry). Soliciting the participation of the development community in the policy-formation process could facilitate smoother implementation and help officials craft policy that is sensitive to advances in management practices and technology.

Woodstock's public square provides a timeless reminder of efficient planning in an expanding suburbia. *(Credit: Charles Hanlon)*

One idea worthy of further consideration is the establishment of a joint council of developers, real estate professionals, and planning officials to discuss problems associated with contemporary development control practices. Such an entity, meeting periodically, could help lessen the level of mistrust and suspicion among stakeholders that can result in misunderstanding or missed opportunities. This coalition could also help alleviate the "us versus them" mentality that often pervades the negotiating process.

On a small scale, municipalities could establish more productive relationships with developers by preparing a "development manual" that summarizes the key aspects of the municipal code and the process employed for project approval. Such a manual might consist of only a few pages listing Internet resources and phone numbers, as well as a short description of municipal development control. Too often, developers are left to sort through much irrelevant information simply to understand the more salient points about subdivision regulation, appearance and architectural review, building permits, and other issues.

Input from developers could augment the approval process for municipalities and other units of government.

At present, the approval processes used by local governments seemingly differ in unpredictable ways. These processes sometimes stifle development projects—and ultimately place key players in adversarial roles. Incremental steps toward improved approval methods, perhaps based on the results of formal conversations between chambers of commerce, neighborhood organizations, economic development groups, and the Urban Land Institute, could produce more efficient practices. Insights from the developer and other real estate professionals also could add to the municipality's sophistication in comprehensive planning, resulting in more thoughtful development.

Local chambers of commerce and industrial and trade associations are particularly logical candidates for establishing new lines of communication.

Encouraging dialogue between stakeholders could help avoid the trial-and-error approach that often characterizes the evolution of municipal approval processes. Such interaction will only grow in importance as the pace of global commerce quickens, rendering successful economic development more dependent on rapid decision-making than ever before.

Time may have brought advances in information and development technology since Chaddick made his observations. Still, the essence of his ideas remains relevant to all professionals with a commitment to equitable and efficient development.

Select References

Altshuler, Alan, and Jose Gomez-Ibanez. 1993. *Regulation for Revenue: The Political Economy of Land Use Exactions.* Cambridge, Mass: The Lincoln Institute of Land Policy.

American Planning Association. 1996. *Growing Smart[SM] Legislative Handbook.* Chicago: APA Press.

Berke, Phillip, and Steven French. 1994. "The Influence of State Planning Mandates on Local Plan Quality." *Journal of Planning Education and Research* 13: 237–250.

Bluestone, Daniel. 1993. *Constructing Chicago.* New Haven: Yale University Press.

Brueckner, Jan. 1999. *Urban Sprawl: Diagnosis and Remedies.* Urbana-Champaign, Ill.: Institute of Government and Public Affairs, University of Illinois at Urbana-Champaign.

Burnham, Daniel, and Edward Bennett. 1993. *Plan of Chicago.* Edited by Charles Moore. New York: Princeton Architectural Press. Original edition: Chicago, Commercial Club of Chicago, 1909.

Byrum, Oliver. 1992. *Old Problems in New Times: Urban Strategies for the 1990s.* Chicago: APA Planners Press.

Calthorpe, Peter. 1993. *The Next American Metropolis.* Princeton, N.J.: Princeton Architectural Press.

Cervero, Robert. 1991. "Land Uses and Travel at Suburban Activity Centers." *Transportation Quarterly* 45, 4 (October): 479–491.

Chaddick, Harry. 1990. *Chaddick: Success Against All Odds.* Chicago: Harry F. Chaddick Associates.

Chapin, Stuart, Jr., and Edward Kaiser. 1979. *Urban Land Use Planning.* 3rd edition. Urbana, Ill.: The University of Illinois Press.

Chicago Area Transportation Study. 1998. *Destination 2020.*

Chicago Central Area Committee and the City of Chicago. 1984. *Chicago Central Area Plan.*

Chicago Department of Public Works. 1973. *Chicago Public Works: A History.*

Chicago Plan Commission. 1973. *Chicago 21.*

City of Chicago. 1995. *Brownfields Forum: Recycling Land for Chicago's Future. Final Report and Action Plan.*

Commercial Club of Chicago. 1998. *Chicago Metropolis 2020: Preparing Metropolitan Chicago for the 21st Century.*

Condit, Carl. 1974. *Chicago 1930–1970: Building, Planning and Urban Technology.* Chicago: The University of Chicago Press.

Cronon, William. 1991. *Nature's Metropolis: Chicago and the Great West.* New York: W.W. Norton and Co.

Cullingworth, Barry. 1993. *The Political Culture of Planning: American Land-Use Planning in Comparative Perspective.* New York: Routledge.

Downs, Anthony. 1994. *New Visions for Metropolitan America.* Washington, D.C.: Brookings Institution and Lincoln Institute of Land Policy.

Environmental Protection Agency. 1996. *Indicators of the Environmental Impacts of Transportation: Highway, Rail, Aviation, and Maritime Transport.* Washington, D.C.: EPA 230-R-96-009.

Fischel, William. 1985. *The Economics of Zoning Laws: A Property Rights Approach to American Land Use Controls.* Baltimore: The Johns Hopkins University Press.

Fischer, Paul. 1999. *Section 8 and the Public Housing Revolution: Where Will the Families Go?* Chicago: Woods Fund.

Garreau, Joel. 1991. *Edge City: Life on the New Frontier.* New York: Doubleday.

Gore, Samuel, and James Nowlan. 1996. *Illinois Politics and Government: The Expanding Metropolitan Frontier.* Lincoln, Neb.: The University of Nebraska Press.

Government Accounting Office. 1995. *Report to the Chair, Committee on Small Business, House of Representatives: Community Development: Reuse of Urban Industrial Sites.* Washington, D.C.: GAO/RCED-95-172.

———. 1996. *Superfund: Barriers to Brownfield Redevelopment.* Washington, D.C.: GAO/RCED-96-125.

Grady, Graham, and Pamela Freese. 1997. *Towards a Regional Plan of Chicago: Shaping a Burnham 2000 Initiative.* Chicago: Chaddick Institute for Metropolitan Development.

Harris, Britton, and Michael Batty. 1993. "Locational Models, Geographic Information and Planning Support Systems." *Journal of Planning Education and Research* 12,3: 184–198.

Hiss, Tony. 1990. *The Experience of Place.* New York: Random House.

Innes, Judith. 1996. "Planning through Consensus Building: A New View of the Comprehensive Planning Ideal." *Journal of the American Planning Association* 62 (Spring): 460–472.

Jacobs, Jane. 1961. *The Death and Life of Great American Cities.* New York: Random House.

Joseph, Lawrence B. 1994. *Affordable Housing and Public Policy.* Chicago: Chicago Assembly.

Kaiser, Edward, and David Godschalk. 1995. "Twentieth Century Land Use Planning: A Stalwart Family Tree." *Journal of the American Planning Association* 61 (Summer): 365–385.

Kamin, Blair. 1998–1999. "Reshaping the Lakefront." *Chicago Tribune,* article series.

Katz, Peter. 1994. *The New Urbanism: Toward an Architecture of Community.* New York: McGraw-Hill, Inc.

Mayer, Harold, and Richard Wade (contributor). 1973. *Chicago: Growth of a Metropolis.* Chicago: The University of Chicago Press.

Metropolitan Planning Council. 1995. *Housing for a Competitive Region.*

———. 1998. *Growing Sensibly: A Guidebook of Best Development Practices in the Chicago Region.*

Miller, Donald. 1996. *City of the Century: The Epic of Chicago and the Making of America.* New York: Simon and Schuster.

Miller, Luther. 1997. "Grade-crossing Safety: Lessons from Fox River Grove." *Railway Age* 198 (March): 47–50.

Northeastern Illinois Planning Commission. 1992. *Strategic Plan for Land Resource Management.*

———. 1997. *Population, Household and Employment Forecasts for Northeast Illinois, 1999–2020.*

———. 1998. *Toward 2020: A Regional Growth Strategy.*

Openlands Project. 1999. *Under Pressure: Land Consumption in the Chicago Region, 1998–2028.*

Orfield, Myron. 1997. *Metropolitics: A Regional Agenda for Community and Stability.* Cambridge, Mass.: The Lincoln Institute of Land Policy, and Washington, D.C.: Brookings Institution.

Regional Transportation Authority. 1995. *The Market for Transit-Oriented Development.*

Rusk, David. 1993. *Cities Without Suburbs.* Washington, D.C.: The Woodrow Wilson Center.

Schlereth, Thomas. 1981. "Burnham's Plan and Moody's Manual." *Journal of the American Planning Association* 47,1: 70–82.

Simmons, Robert. 1998. *Turning Brownfields into Greenbacks.* Washington, D.C.: The Urban Land Institute.

So, Frank, and Judith Getzels, eds. 1988. *The Practice of Local Government Planning,* 2nd edition. Washington, D.C.: International City Management Administration.

Stone, Deborah C. 1995. *Creating a Regional Community: The Case for Regional Cooperation.* Chicago: Metropolitan Planning Council.

Sustain. 1999. *Beyond Sprawl: A Guide to Land Use in the Chicago Region for Reporters and Policy Makers.*

Suttles, Gerald D. 1990. *The Man-Made City: The Land-Use Confidence Game in Chicago.* Chicago: The University of Chicago Press.

Teitz, Michael. 1996. "American Planning in the 1990s: Evolution, Debate and Challenge." *Urban Studies* 33: 649–671.

Texas Transportation Institute. 1998. *Urban Roadway Congestion, 1982–1996.*

Tiebout, Charles. 1956. "A Pure Theory of Local Expenditures." *Journal of Political Economy* 64 (February): 416–424.

ULI–the Urban Land Institute. 1996. *New Uses for Obsolete Buildings.* Washington, D.C.: The Urban Land Institute.

———. 1997. *Creating More Livable Metropolitan Areas.* Washington, D.C.: The Urban Land Institute.

University of Illinois at Chicago, Urban Transportation Center. 1998. *Highways and Urban Decentralization.*

U.S. Conference of Mayors. 1998. *Recycling America's Land: A National Report on Brownfield Redevelopment.*

Wei, Ge. 1995. "The Urban Enterprise Zone." *Journal of Regional Science* 35,2: 217–231.

Wiewel, Wim, and Phillip Nyden, eds. 1991. *Challenging Uneven Development: An Urban Agenda for the 1990s.* New Brunswick, N.J.: Rutgers University Press.

Wille, Lois. 1997. *At Home in the Loop: How Clout and Community Built Chicago's Dearborn Park.* Carbondale, Ill.: Southern Illinois University Press.

Wille, Lois, and Gerald Suttles (designer). 1991. *Forever Open Clear and Free: The Struggle for Chicago's Lakefront.* Chicago: University of Chicago Press.

Willis, Carol. 1997. *Form Follows Finance: Skyscrapers and Skylines in New York and Chicago.* New York: Princeton Architectural Press.

Index

About the Author

Martin E. Toth holds a master of science degree in public service management from DePaul University and a bachelor's degree from the University of Colorado. He is program director at the Chaddick Institute for Metropolitan Development at DePaul University in Chicago. Originally from Long Beach, Calif., he served as a Peace Corps volunteer in Central America and has conducted extensive research on planning and development control under the auspices of the Chaddick Institute.

About the Editor

Joseph P. Schwieterman, Ph.D., is an associate professor of Public Services Management at DePaul University and director of the school's Chaddick Institute for Metropolitan Development. He holds a master of science degree in transportation from Northwestern University and a doctoral degree in public policy studies from the University of Chicago. A long-standing contributor to the Transportation Research Board, Professor Schwieterman has published extensively on the evolution of urban and suburban land use and transportation.

About the Chaddick Institute

Since its creation in 1993, the Chaddick Institute for Metropolitan Development, located at DePaul University in Chicago, has advanced the principles of effective land-use, transportation, and infrastructure planning. The Institute offers planners, attorneys, developers, and entrepreneurs a forum to share expertise on difficult land-use issues through workshops, conferences, and policy studies.

Over the past year, the Chaddick Institute has offered educational events in cooperation with the Illinois chapter of the American Planning Association, Lambda Alpha International, Illinois Rail, Scenic Illinois, Women in Planning and Development, and other organizations. It provides student scholarships and offers a three-course Certificate in Metropolitan Planning in cooperation with the Public Services Graduate Program.

Financial support for the Chaddick Institute is provided by the Harry F. and Elaine M. Chaddick Foundation. Additional information about the Institute and a summary of the civic contributions of Mr. Chaddick are available at its Web site at www.depaul.edu/~chaddick.